Lecture Notes in Bioengineering

Lecture Notes in Bioengineering (LNBE) publishes the latest developments in bioengineering. It covers a wide range of topics, including (but not limited to):

- Bio-inspired Technology & Biomimetics
- Biosensors
- Bionanomaterials
- Biomedical Instrumentation
- Biological Signal Processing
- Medical Robotics and Assistive Technology
- Computational Medicine, Computational Pharmacology and Computational Biology
- Personalized Medicine
- Data Analysis in Bioengineering
- Neuroengineering
- Bioengineering Ethics

Original research reported in proceedings and edited books are at the core of LNBE. Monographs presenting cutting-edge findings, new perspectives on classical fields or reviewing the state-of-the art in a certain subfield of bioengineering may exceptionally be considered for publication. Alternatively, they may be redirected to more specific book series. The series' target audience includes advanced level students, researchers, and industry professionals working at the forefront of their fields.

Indexed by SCOPUS, INSPEC, zbMATH, SCImago.

More information about this series at https://link.springer.com/bookseries/11564

Borja Sañudo Corrales ·
Jerónimo García-Fernández
Editors

Innovation in Physical Activity and Sport

Selected Papers from the 1st International Virtual Conference on Technology in Physical Activity and Sport

Springer

Editors
Borja Sañudo Corrales
Department of Physical Education and Sport
University of Seville
Sevilla, Sevilla, Spain

Jerónimo García-Fernández
Department of Physical Education and Sport
University of Seville
Sevilla, Sevilla, Spain

ISSN 2195-271X ISSN 2195-2728 (electronic)
Lecture Notes in Bioengineering
ISBN 978-3-030-92896-4 ISBN 978-3-030-92897-1 (eBook)
https://doi.org/10.1007/978-3-030-92897-1

This Springer imprint is published by the registered company Springer Nature Switzerland AG
The registered company address is: Gewerbestrasse 11, 6330 Cham, Switzerland

Preface

The number of new consumer digital technologies is growing exponentially, and today it is almost impossible to disassociate them from our lifestyle. The enormous interest in these types of devices has led to wearable technologies being rated as the primary trend in fitness worldwide with considerable increases in sales predicted over the coming decade. Interest among researchers is also growing exponentially as judged by the number of studies being published on consumer digital technologies. This book comprises selected peer-reviewed papers presented at the International Conference on Technology in Physical Education and Sport (TAPAS 2020) organized by University of Seville, Spain. The book covers latest research on a wide range of technologies used in the assessment of physical fitness and health, for biomechanical analysis and for sport education and rehabilitation.

In the first part of the book, a wide range of technologies including Global Positioning Systems and portable wearable metabolic devices are discussed. These technologies can provide information of the internal physiological and mechanical loading demands of athletes in different sports. Moreover, they help gaining a deeper understanding of athlete's and team's performance.

In the second part, innovations in the field of sport and health sciences are discussed. Emerging applications of tools such as exergames, and virtual and augmented reality are described. A set of pilot studies and systematic reviews have been collected in this part to provide researchers with background information and ideas for future studies.

Overall, this book should help bridge the gap between laboratory research and professional training, highlighting current technologies that can be used and developed further to enhance performance or improve functionality in different population groups. It aimed at providing a practical, evidence-based vision of how to use technologies, addressing sport scientists, coaches and physiotherapists alike.

Borja Sañudo Corrales
Jerónimo García-Fernández

Contents

About the Editors

Borja Sañudo Corrales
bsancor@us.es
Physical Education and Sport Department, Universidad de Sevilla
C/ Pirotecnia, s/n, 41013, Seville, Spain

He received the B.S. and M.S. degrees in physical activity and health from the International University of Andalusia in 2007 and the Ph.D. degree in sport science from University of Seville, Spain, in 2009. From 2010, Borja is Full Professor at the Department of Physical Education and Sport (University of Seville). His major research interests are centred on the role of exercise and other lifestyle factors for promoting improvements in physiological function, quality of life and disease-free survival in clinical populations and improvements in athletes′ performance. Since 2005, he has been working on a number of studies with clinical populations (mainly fibromyalgia, Type 2 diabetes and obesity) and athletes (whole body vibration and exercise and the use of technologies in sports). He is Author of five books, more than 100 articles in peer-reviewed journals and more than 80 international presentations.

Jerónimo García-Fernández
jeronimo@us.es
Physical Education and Sport Department, Universidad de Sevilla
C/ Pirotecnia, s/n, 41013, Seville (Spain)

Associate Professor in the Physical Education and Sport Department at the University of Seville, Spain, where he teaches Bachelor and Master-level courses in sports management and marketing. He has published more than 80 articles indexed in international journals as well as various chapters. His main research topics are sport management, sport technology, fitness apps, customer loyalty and satisfaction at fitness services.

Technologies Applied to Enhance Performance in Sports

Analysis of Explosive Force, Sprint Distance and High-Intensity Running in a Match Situation Between Hungarian Second-Division Soccer Players

Soós Imre[1(✉)], Gyagya Attila[2], Kósa Lili[2,3], K. J. Finn[4], and Ihász Ferenc[1,2]

[1] Doctoral School of Health Sciences, University of Pécs, Pécs, Hungary
imre.soos@eto.hu
[2] ETO FC, Győr, Hungary
[3] Eötvös Lóránd University Psychology Doctoral School, Budapest, Hungary
[4] School of Nutrition, Kinesiology, and Psychological Sciences College of Health, Science and Technology, University of Central Missouri, Warrensburg, USA

Abstract. Match play competitive in team sports requires players to perform actions resulting in frequent intense acceleration and deceleration. Accordingly, these accelerations and decelerations make up a substantial part of the high-intensity external workload, imposing distinctive and disparate internal physiological and mechanical loading demands on players. The aim of the present study is to compare the difference between the load characteristics of five league matches. Twenty-eight ($n = 28$); (age = 22.0 years ± 5.9) young professional players participating at least five games during their examined period. The observation of the players took place in the five weeks following the changes in their training, which also meant a fundamental change in training stimuli. Data was collected through electronic performance tracking systems ("Catapult Vector").The combined team-level results [explosive efforts (EE), sprint distance (SD) and high-intensity running (HIR) total player load (TPL)] of the five matches and their weekly differences were compared with the Repeated measures ANOVA with Tukey HSD post hoc method, the level of significance was set at $p \leq 0.05$.The characteristics of the three matches following the coach change showed a significant difference. Sprint distance (SD) and high-intensity running (HIR) shows a significant difference between (1–2; 1–3; 1–4); ($p < 0.001$) matches. The same is true for team-level averages of total player load (TPL); ($p < 0.05$). The findings of this study provide meaningful information regarding explosive efforts (EE), sprint distance (SD), high-intensity running (HIR) as well as total player load (TPL) profiles of professional football matches.

Keywords: Team sports · Total player load · Team-level · Mechanical loading

B. Sañudo Corrales and J. García-Fernández (Eds.): Tapasconference 2020, LNBE, pp. 3–8, 2022.
https://doi.org/10.1007/978-3-030-92897-1_1

1 Introduction

Athlete monitoring has become a contemporary research topic as well as a service, which has benefited an athlete's progress in training and during competition [1]. Several monitoring systems are on the market, which allows the sport trainers to determine loads on the athlete [2]. Football is one sport that requires high-intensity exertion based upon the movements of the athletes during play [3, 4]. Measures of running performance during matches identify that professional football players run more than 10 km however only 10% of this distance is at a high velocity [3, 5]. Unfortunately, these intermittent spurts of high-intensity efforts can contribute to injury due to fatigue [6]. The football player accelerates their run speed during play and often decelerates which challenges the athlete's neuromuscular function and metabolic capacity [7–10]. Player loads based upon external loads with these changes in run velocity had been studied in elite soccer match [11] using monitoring systems which provide evidence that profiles of football players with high velocity (19.8–25.2 km. hr -1) have the highest physical demands. These demands are evident when evaluating the perception of effort during training sessions [12]. More investigations in this area will allow sport trainers to help prevent injury and optimize training which will influence the performance of football players during competition.

The study aim was to compare the combined team-level profile differences in the high intensity running (HIR), explosive efforts (EE) and sprint distance (SD) of Hungarian professional football players. This study was conducted over eight competitive weekly periods in a professional football team from the second divisions.

1.1 Subjects

Twenty-eight young (age = 22.0 years ± 5.9) professional players participating at least five games during their examination period. The club allowed the research team to access players' data and informed consent was provided. The study was conducted ethically according to Declaration of Helsinki and it was approved by the Bioethics Committee at the University of Pécs. Anthropometric measures were taken at the beginning of the study period. The averages (± sd) of the assessments for body mass was 78.7 ± 8 kg, body height was 181.7 ± 10.1 cm, and percent body fat 10.2 ± 2.3%.

2 Materials and Methods

2.1 Study Design

The observation of the players took place in the eight competitive weekly period he changes in their training, which also meant a fundamental change in training stimuli. The match location (home or away) alternated with each macrocycle; three matches were played away and two at home. The playing formation was 4–3-3 for all matches. After the first match, there was a substitution of coach and assistant coach as the previous period proved unsuccessful.

Data was collected through electronic performance tracking systems using the Catapult Vector GPS system. This device contains inertial sensors (four 3D accelerometers, three 3D gyroscopes, one 3D magnetometer and one barometer), which collected data at 100 Hz. The validity and reliability of this device has been analyzed for the collection of time-motion variables and is considered a suitable instrument for this purpose in football. The five matches and their weekly differences of the combined team-level results of explosive efforts (EE), sprint distance (SD) and high-intensity running (HIR) total player load (TPL) were compared. Using a repeated measures Analysis of Variance (ANOVA) with the Tukey HSD post hoc method, the level of significance was set at $p \leq 0.05$.

2.2 Training Stimuli

The new professional staff introduced a new content and form of training work: The main difference between the two coaches was the modality of approaching the game. The Coach that was ultimately changed preferred a more tactical and ball possession-based approach of the match as the other Coach had more emphasis on the physical side. Because it is very difficult to develop the physical component of a team during the season, we decided that we would follow the Newcastle Utd way of developing the stamina of a player presented by John Fitzpatrick, et al. 2018 [13]. In essence, that means that a player has to spend more than 6.30 min above their individual maximal aerobic speed (MAS) during a weekly macrocycle to preserve their stamina and above 9 min to develop. If this goal was not achieved during ball-based games and drills than interval running was introduced [14].

3 Results

Figure 1 Illustrates the changes in sprint distance and high intensity running. Going from left to right from the black vertical line, the team averages of the two characteristics increase. From the third match the pattern of the two characteristics changes. For sprint distance (SD), a significant ($p < 0.05$) difference between the second match (58.7 ± 15.3 min) and matches three (126.6 ± 36 min). For High Intensity Running (HIR), significant increases from matches three to five were observed (383.5 ± 121.4 vs. 497.7 ± 124.6 min).

Abbreviation: Sprint Distance (>25.2 km/h), HIR = High Intensity Running (19.8–25.2 km/h), * = $p < 0.05$, 1, 2, 3, 4, 5

The numbers on the horizontal axis shows five matches, from the black vertical line begun the matches with new Coach, and new assistant Coach.

With explosive strength (EE), a significant difference between matches 2–5 were evident (Fig. 2). Each match showed increasing EE. The EE for match two was 19.7 ± 7.4 min vs. 22.5 ± 9.7 min for match three. Match four increased to 24.1 ± 10.5 min and match five to 32.3 ± 10.6 min. For the total player load (TPL), we found only a significant difference between the matches two (702.4 ± 259.4) and five (885.4 ± 272.4).

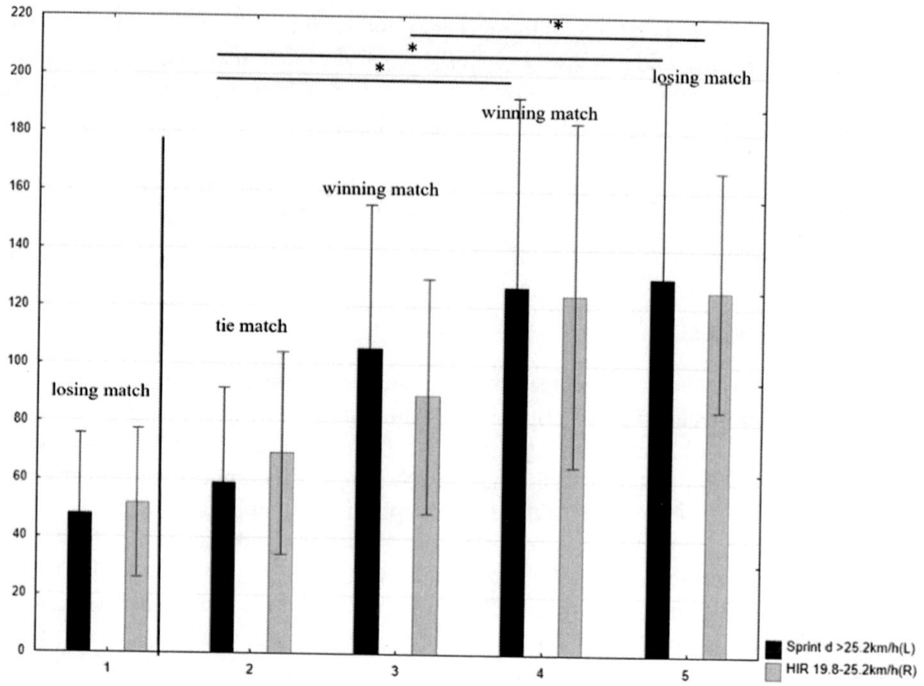

Fig. 1. Changes of the Sprint distance (>25.2 km/h), and High Intensity Running (19.8–25.2 km/h) during the monitored five matches

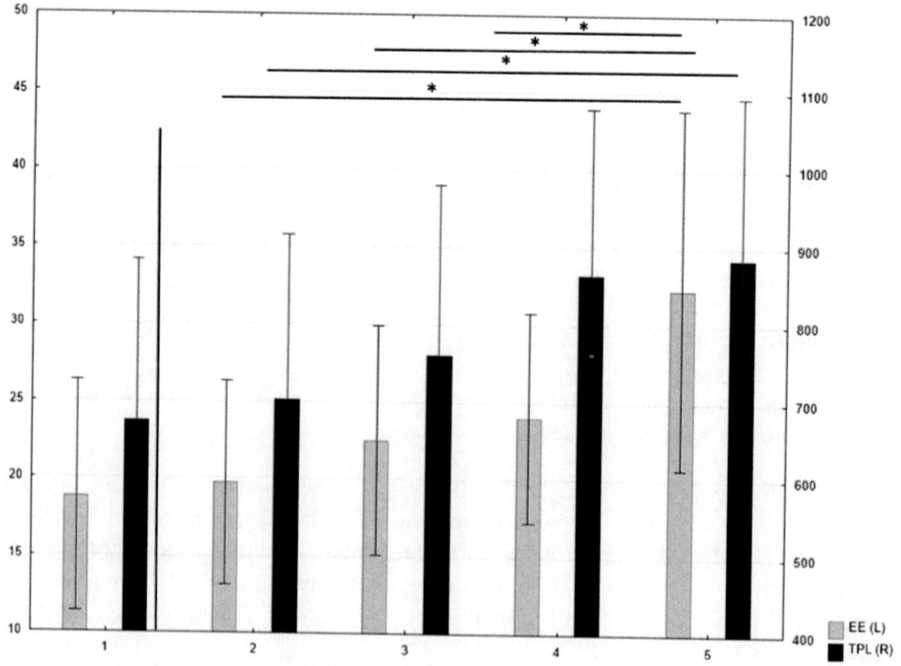

Fig. 2. Changes of the Explosive Efforts, and Total Player Load during the monitored five matches

Abbreviation: EE = Explosive Efforts, TPL = Total Player Load $p < 0.05$, 1,2,3,4,5

The numbers on the horizontal axis shows five matches, from the black vertical line begun the matches with new Coach, and new assistant Coach

4 Discussion

The purpose of this study was to examine the difference in a team environment of the physiological output between championship games in a transition of two different approach of coaching style. In terms of performance, the team lost in the first match, a draw in the second, was the third and fourth winning match, and lost again in the fifth (Fig. 1).

If we compare the team averages from the five matches, we can assume that the parameters examined here are not direct indicators of success at all times. This means, that the tactical environment (e.g. team tactics, opposing team's tactics, game dynamics) is presumably shaping the physiological needs of a given match. Going from match 1–3, we can observe a slight increase in both physiological performance and success. Match 4 required even more high-speed running. Regarding the measured parameters, match 5 is similar to match 4, although a significant increase in explosive efforts is present. We don't seem to be able to say exactly what kind of proportion appears the physiological performance, technical, or tactical elements during the mentioned matches.

The findings of this longitudinal study provide meaningful information regarding explosive efforts (EE), sprint distance (SD), high-intensity running (HIR) as well as total player load (TPL) profiles of professional football matches. This data could also serve as a comparison source for future researchers or sports scientists and coaches from professional football teams.

References

1. Gabbett, T., et al.: The athlete monitoring cycle: a practical guide to interpreting and applying training monitoring data. Br. J. Sports Med. **51**(20), 1451–1452 (2017). https://doi.org/10.1136/bjsports-2016-097298. (pmid:28646100)
2. Gómez-Carmona, C.D., Pino-Ortega, J., Sánchez-Ureña, B., Ibáñez, S.J., Rojas-Valverde, D.: Accelerometry-based external load Indicators in sport: too many options, same practical outcome? Int. J. Environ. Res. Public Health **16**(24), 5101 (2019). https://doi.org/10.3390/ijerph16245101. (pmid:31847248)
3. Rampinini, E., Coutts, A., Castagna, C., Sassi, R., Impellizzeri, F.: Variation in top level soccer match performance. Int. J. Sports Med. **28**(12), 1018–1024 (2007). https://doi.org/10.1055/s-2007-965158. (pmid:17497575)
4. Oliva-Lozano, J.M., Gómez-Carmona, C.D., Pino-Ortega, J., Moreno-Pérez, V., Rodríguez-Pérez, M.A.: Match and training high intensity activity-demands profile during a competitive mesocycle in youth elite soccer players. J. Human Kinet. **75**(1), 195–205 (2020). https://doi.org/10.2478/hukin-2020-0050
5. Palucci, L., et al.: Running performance in Brazilian professional football players during a congested match schedule. J. Strength Cond. Res. **32**(2), 313–325 (2018). https://doi.org/10.1519/JSC.0000000000002342. (pmid:29369952)

6. Carling, C., Gall, F.L., Reilly, T.P.: Effects of physical efforts on injury in elite soccer. Int. J. Sports Med. **31**(03), 180–185 (2010). https://doi.org/10.1055/s-0029-1241212. (pmid:20024885)
7. Buchheit, M., et al.: Mechanical determinants of acceleration and maximal sprinting speed in highly trained young soccer players. J. Sports Sci. **32**(20), 1906–1913 (2014). https://doi.org/10.1080/02640414.2014.965191. (pmid:25356503)
8. Haugen, T.A., Tønnessen, E., Seiler, S.: Anaerobic performance testing of professional soccer players 1995–2010. Int. J. Sports Physiol. Perform. **8**(2), 148–156 (2013). https://doi.org/10.1123/ijspp.8.2.148. (pmid:22868347)
9. Harper, D.J., Carling, C., Kiely, J.: High-intensity acceleration and deceleration demands in elite team sports competitive match play: a systematic review and meta-analysis of observational studies. Sports Medicine **49**(12), 1923–1947 (2019). https://doi.org/10.1007/s40279-019-01170-1. (pmid:31506901)
10. Hader, K., Mendez-Villanueva, A., Palazzi, D., Ahmaidi, S., Buchheit, M.: Metabolic power requirement of change of direction speed in young soccer players: not all is what it seems. PLOS ONE **11**(3), e0149839 (2016). https://doi.org/10.1371/journal.pone.0149839. (pmid:26930649)
11. Dalen, T., Jørgen, I., Gertjan, E., Havard, H.G., Ulrik, W.: Player load, acceleration, and deceleration during forty-five competitive matches of elite soccer. J. Strength Cond. Res. **30**(2), 351–359 (2016). https://doi.org/10.1519/JSC.0000000000001063. (pmid:26057190)
12. Paolo Gaudino, F., Iaia, M., Strudwick, A.J., Hawkins, R.D., Alberti, G., Atkinson, G., Gregson, W.: Factors influencing perception of effort (session rating of perceived exertion) during elite soccer training. Int. J. Sports Physiol. Perform. **10**(7), 860–864 (2015). https://doi.org/10.1123/ijspp.2014-05184. (pmid:25671338)
13. Fitzpatrick, J.F., Hicks, K.M., Hayes, P.R.: Dose–response relationship between training load and changes in aerobic fitness in professional youth soccer players. Int. J. Sports Physiol. Perform. **13**(10), 1365–1370 (2018). https://doi.org/10.1123/ijspp.2017-0843
14. Buchheit, M.: The 30–15 Intermittent Fitness Test: 10 year review it Myorobie, 1 September 2010. http://www.martin-buchheit.net

Applications and Efficacy of Portable Wearable Metabolic Devices

Eric Gasmin[1,2], Leslie Yessenia Castillo-Ortiz[1,2], Ryan P. Durk[1,2], Kent A. Lorenz[1,2], Marialice Kern[1,2], C. Matthew Lee[1,3], and James R. Bagley[1,2,3](✉)

[1] Exercise Physiology Laboratory, Department of Kinesiology, San Francisco State University, San Francisco, CA, USA
jrbagley@sfsu.edu

[2] Virtual Reality Institute of Health and Exercise, San Francisco, CA, USA

[3] Healthy Living for Pandemic Event Protection (HL-PIVOT) Network, San Francisco, CA, USA

Abstract. Metabolic carts use indirect calorimetry to measure energy expenditure (EE). Because these systems are large, portable wearable metabolic devices (PWMD) have been developed to measure EE outside of the laboratory. The purpose of this review article is to provide an overview of the current state of knowledge regarding the 1) application and 2) efficacy of portable metabolic devices. This literature review covered the applications and efficacy of PWMD during exercise. Studies that investigated the efficacy of PWMD evaluated the volume of oxygen consumption (VO_2) and carbon dioxide production (VCO_2) measurements compared to criterion devices. Thirty-four articles studied the application and efficacy of PWMD during walking, running, cycling, rowing, badminton, and tennis. Eight walking and running studies reported that PWMD overestimated VO_2 (0.30–0.63 L/min; ~5.2–22.2% difference) and VCO_2 (0.42–0.62 L/min; ~3% difference) compared to criterion devices. Nine cycling studies showed similar trends for VO_2 (0.1–0.3 L/min; ~4.4–13.0% difference) and VCO_2 (0.14–0.30 L/min; 6.9% difference). Generally, as exercise intensity increased, variability between devices correspondingly increased. However, six studies reported no significant differences between PWMD and criterion devices (ICC: 0.93–0.97; r = 0.91–.99). Regarding reliability, five studies reported that popular PWMD had high test-retest reliability for VO_2 (ICC: 0.87–0.99) and VCO_2 (ICC: 0.72–0.81) while walking/running at different speeds/gradients. Conversely, one study reported low test-retest reliability while running at high speeds for short distances. As technology emerges, PWMD are becoming widely used in clinical, fitness, and research. While data on their validity remains ambiguous, their test-retest reliability may be a more important factor to consider when using a PWMD.

Keywords: Metabolic cart · Cardiorespiratory fitness · Validity · Reliability · VO_2

E. Gasmin and L. Y. Castillo-Ortiz—Co-first Authorship

B. Sañudo Corrales and J. García-Fernández (Eds.): Tapasconference 2020, LNBE, pp. 9–25, 2022.
https://doi.org/10.1007/978-3-030-92897-1_2

1 Introduction

Researchers have conventionally used stationary analyzers to measure human metabolic function. Maximal oxygen consumption (VO_2max), commonly assessed through graded exercise tests (GXT), is the gold standard measure of an individual's maximal aerobic capacity and cardiorespiratory fitness (CRF) [1]. Regularly assessing VO_2max best predicts overall fitness related to performance [2] as it can track an individual's training adaptations [3].

Cardiopulmonary exercise testing (CPET) is a reliable tool to assess exercise performance in athletic and clinical settings [4]. CPET usually takes place in controlled laboratory environments with participants placed on treadmills or cycle ergometers whilst expired air is simultaneously measured via gas analyzers [5]. These analyzers can measure metabolic variables in two ways: 1) collecting and analyzing expired air in mixing chambers or 2) measuring real-time expiration with mechanoelectrical turbines [6–8]. Examples of previously validated metabolic carts include the CosMed Quark CPET and ParvoMedics TrueOne 2400 [5, 9–15]. Despite their accuracy, these metabolic carts are large, cumbersome, and limited to a physical space. Additionally, CPET requires trained personnel, which can be costly, and usually limited to research or patients for clinical diagnosis and prognosis [16].

To breach physical limitations and increase availability, researchers and engineers have developed portable wearable metabolic devices (PWMD) aiming to accurately measure real-time metabolism in natural field settings [9, 17, 18]. Device portability can allow users to monitor real-time CRF, as well as prescribe personalized exercise programs [19, 20]. To know an individual's aerobic capacity, these analyzers must assess volume of oxygen consumption (VO_2) and carbon dioxide production (VCO_2) [19, 21, 22]. Common PWMD include the Oxylog, TEEM 100, Cortex X1, SensorMedics VmaxST, Oxycon Mobile Metabolic System, Metamax I/II, and Cosmed (K2, K4, K4b^2, and K5). This literature review provides a current overview regarding the 1) applications and 2) efficacy (validity and reliability) of these PWMD. The studies discussed here compared devices to various criterion measures, in different settings, and during diverse exercise protocols.

2 Methods

We performed a systematic search of the peer-reviewed literature on PWMD use during exercise or physical activity. Various databases and search engines (i.e., Google Scholar, NCBI, Sportdiscus) were used with the following keywords: "portable metabolic analyzers", "portable metabolic analysis", "portable VO2", "validation of portable devices", "portable exercise metabolism", "validity and reliability of portable metabolic devices". Articles were acquired from 1982 to 2020 (accessed between August 15 to October 15, 2020). Throughout the search, studies were considered for the application and efficacy of PWMD.

Articles were only included in this review if they met the following criteria: 1) A potable metabolic device must have been used during any form of physical activity, regardless of environment, intensity, frequency, duration, and volume of exercise and 2) efficacy studies included validity (accuracy) and/or reliability (test-retest) components. Validation studies observed the accuracy of PWMD compared to a criterion measure. Regardless of which criterion device was used for validation, the study was included. Studies were also considered despite their findings being significantly different (p-value) or having no significant differences, expressed with intra-class correlations (ICCs) or standard error of measurements (SEM). Test-retest reliability used the same PWMD across different time periods. Other forms of reliability were not considered.

3 Results

Thirty-four articles met the criteria. Thirteen studies used PWMD while cycling, 11 running, five walking, two mimicked swimming, one playing badminton [23], one playing tennis [24], and one investigated both soccer and rowing modalities [25] (Fig. 1).

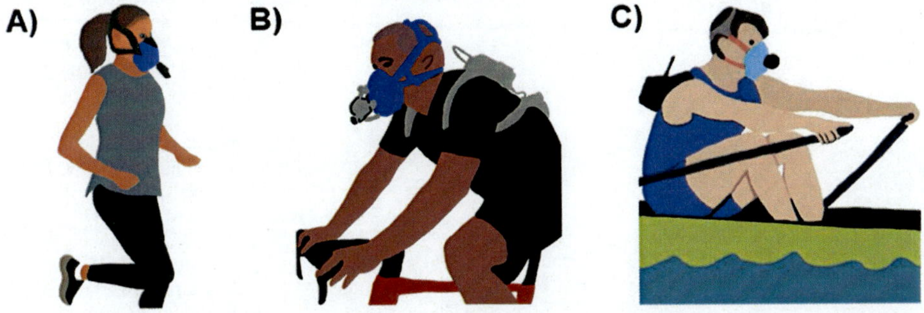

Fig. 1. Exercise modalities performed using portable wearable metabolic devices (PWMD) can include A) walking or running, B) cycling, and C) rowing.

3.1 Applications

We identified eight applications for PWMD used in the literature. Table 1 shows the application, device used, and exercise protocol. Walking and running studies altered intensity by modifying speeds (m/s) and gradients (%) [26]. Studies took place indoors on a treadmill [11, 27, 28] or a combination of indoors and outdoors [29]. Cycling studies took place indoors on an ergometer, but varied intensity by Wattage (W) or kilopond (kp) [25]. Exercise protocols differed in Wattage increments, via Watt maximum percentage (Wmax%) [30], 25 W [31], or 50 W [13, 32, 33].

3.2 Efficacy: Validity and Test-Retest Reliability

Table 2 shows efficacy studies organized by application highlighting the PWMD device, criterion measure, protocol, and validity and/or reliability results. Generally, most studies only reported a validation component, while eight studies performed both validity and reliability. Validation studies recorded significant differences ($p \leq 0.05$) between devices with VO_2 and VCO_2 expressed in absolute units (L/min) or percentages (%). Studies that incorporated test-retest reliability recorded intra-class correlations (ICCs), Pearson's correlation (r), or standard error of measurement (SEM).

Table 1. Summary of studies with various applications using portable wearable metabolic devices (PWMD).

Application	Study	Device	BxB or MC	Protocol
Running	Ballal & Macdonald (1982)	Oxylog	DB Method(BxB)	Treadmill: 12min continuous exercise test 0 gradient: 3 min at each speeds (2.9,4.8, 6.4, and 8.0 km.h)
	Wideman et. al. (1996)	Aerosport TEEM 100	Rayfield Metabolic Measuring System	Two Protocols (INC)--horizontal treadmill: increased 10 m·min-1 every 3 min. incremental step protocol: 8-inch step; Step rate increase: 3 min by 2.5 steps·min-1 until exhaustion (CL)-- run:30 min Intensities: 1) 60% HR max, 2) 90% HR max, 3) blood lactate concentration of 4.0 mM
	Melanson et al. (1996)	Aerosport TEEM 100	On-line computer based acquisition metabolic system	Submaximal Exercise: 10 min Slow walking (6.2 km/hr), slow jogging (8.5 km/hr), and running (10.7 km/hr). Speed increased at min 10; 20 GXT: Self-selected speed; 0% grade; increased 2.5% every 2min.
	Kawakami et al. (1992)	K2	DB Method	400m track; increased speed every lap; 6 laps total final test: full speed identical running protocol: 1000m above sea level
	Lothian et al. (1993)	K2	Quinton on-line system	3 continuous protocol (treadmill at 1% gradient) walking: 4km/hr; speed: increased 2km/hr every 3 min until failure

(*continued*)

Table 1. *(continued)*

	Lucia et al. (1993)	K2	DB Method	3 test sessions on treadmill Submaximal: progressive protocol 6 stages of 3min workload: individually determined Maximal: progressive protocol: 1min stages Gradient: increased by 1%/min until failure
	Crandall, Taylor, & Raven (1994)	K2	BxB system	Mechanical lung (10, 60min warmup); Two GXTs: Bruce protocol
	Maiolo et al. (2003)	K4	Airspec QP9000	Incremental running test 8km/h to 18km/h to volitional exhaustion (increase of 2km/h for each level)
	Doyon et al. (2001)	K4b²	Mixing Box System (Hughson et. al., (1980) O_2 Analyzer (Applied Electrochemistry Inc)	running on motorized treadmill speed: increased every 2 min graded increased 2% every 2 min until exhaustion Men: • 2.7 to 3.1 to 3.6 m/s • 6 to 7 to 8 mph Women • 1.8 to 2.2 to 2.7 m/s • 4 to 5 to 6 mph
	Duffield et al. (2004)	K4b²	CO_2 Analyzer (Beckman LB-2) Metabolic Cart: Morgan Ventilometer and Ametek Applied Electrochemistry	4 sessions, separated by 1 day--1week, tested during same time of day Treadmill running at 1% grade with 10 min rest period in between each session • 10 min at 10 km/h • 3 mins at 16km/h • 1 minute 20 km/h
Cycling/ Ergometer	Harrison, Brown, & Belyavin (1982)	Oxylog	Centronics quadrupole mass spectrometer	Cycle ergometer • intensities: 75-150W • work rate: 3-10min each • pedaling rate: 50rpm

(continued)

Table 1. (*continued*)

Carter & Jeukendrup (2002)	OxyCon Alpha OxyCon Pro Pulmolab	DB Method	Four different ventilation simulator levels across 2 days: • VE: 20, 40, 80, and 160 L/min • VO2: equivalent to 1.0, 2.0, 3.0, and 4.0 L/min Three separate visits: • cycled at 100 W • cycled at 150 W 85 minutes of continuous cycling
Crouter et al. (2006)	Med-Graphics VO2000 ParvoMedics True-One 2400	DB method	Cycling 10 min: 50 W, 100 W, 150 W 11 min: 200 W and 250 W Each trial had an 11 min rest period Two trials within 48 h: in randomized fashion ParvoMedics and DB • 5 min at 50, 100, and 150 W • 6 min at 200 and 250 W Equipment switched out and VO2000 connected
Kawakami et al. (1992)	K2	DB Method	Cycling (1 kp) Cycling (1-2.5 kp) Cycling (3-4.5 kp)
Forkink et al. (1994)	K2	Oxyconbeta	Reproducibility of K2 Cycling: three sessions with 2 weeks in between • started at 75 W • increased by 25 W every 3 minutes
Bigard & Guezennec (1995)	K2	Metabolic measure cart (MMC): SensorMedics 2900 (SensorMedics Corp., Yorba Linda, CA)	Cycle Ergometer: 2 testing sessions at different barometric pressures • Sea Level: 759.7 ± 7.8 mmHg • Altitude Chamber simulating 2000m: 591.5 ± 0.5 mmHg) 10 minute warm up

(*continued*)

Table 1. *(continued)*

	Hausswirth et al. (1997)	K4	CPX, Medical Graphics	Increased 25 W every 2 minutes until exhaustion Evaluated at 25%, 50%, and 75% Watt max (Wmax) Cycling Ergometer Two GXT, VO_2max 3 day in between tests • Warm up at 100 W for 6 min • increased by 25 W every 2 minutes Submaximal steady-state cycling • 12 min at 25% Wmax • 12 min at 50% Wmax • 12 min at 75% Wmax
	McNaughton et al. (2005)	K4b^2	Morgan EX670	Maximal & submaximal cycling (150-300W) Protocol: 5 increments, 150W, 200W, 250W, 300W Test 1: Protocol Test 2: started at 150 W for 3 mins, followed by 30 W increments/ 3 minutes to exhaustion.
Walking	Peel & Utsey (1993)	K2	BxB oxygen sensor; No CO_2 sensor	Repeated measures design; Rest (5 mins), Treadmill walking:3 mph Gradients: 0%, 5%, 10%, 15% for 4 mins.
	Schrack et al. (2010)	K4b^2	Medgraphics D Series	Two 400m walk treadmill tests constant, self-selected pace (LDCW test)
	Perez-Suarez et al. (2018)	Cosmed K5	Vyntus CPX metabolic cart (BxB mode)	Prolonged walks (13km) speed at 5km/h
	White, Deblois, & Barreira (2019)	Cosmed K5	Cosmed K4 and K5 in randomized counterbalanced order (BxB mode)	treadmill walking protocol 3 trials: 5-min stages 6 walking speeds 1.5-4.0mph; 0.5mph increments) 2 min rest between stages

(continued)

Table 1. *(continued)*

Swimming	Rodriguez et al. (2007)	$K4b^2$	BxB Swimming snorkels and valve system	Standardized GESS testing, protocols *(designed to mimic swimming conditions)
	Keskinen et al. (2003)	$K4b^2$	BxB newly modified measurement system with swimming snorkel and valve device	Stationary bike ergometer-- stepwise protocol warm-up (70-100W), three 4-min loads (100-150-200W) *(designed to mimic swimming conditions)
Badminton	Rampichini et al. 2018	$K4b^2$	BxB	On-Court: 5-min rally simulations 3 different intensities (low: RPE<10, moderate: RPE<14, and high: RPE<19)
Tennis	Bekraoui et al. (2012)	VmaxST	BxB	Indoor hardcourt single testing session: low speed (4.0-7.0km/h);random order on half-width/ full-width court displacement repeated at high speeds (6.0-9.0km/h)
Soccer	Kawakami et al. (1992)	K2	DB Method	Dribbling, soccer exercises
Rowing	Kawakami et al. (1992)	K2	DB Method	Water course; increase row boat speed: stepwise manner

BxB: Breath by Breath technique; DB Method: Douglas Bag Method; VO_2max: maximal oxygen consumption; GXT: Graded exercise Test; LDCW Test: Long Distance Corridor Walk; INC: Incremental Exercise; CL: Constant load Exercise; MWR: Maximal work rate; RPE: Rate of perceived exertion; HR: Heart rate; m: meters; s: second; W: Watts; km/h: Kilometers per hour; mph:miles per hour; kp: kilopond

Table 2. Summary of portable wearable metabolic device (PWMD) studies investigating validity and reliability (test retest) organized by application

Device	Study	Criterion Measure	Protocol	Validity (L/min or % difference)		Reliability (ICC or SEM)	
				VO_2 (L/min)	VCO_2 (L/min)	VCO_2 (L/min)	VO_2 (L/min)
Cortex XI	Scultz et al., 1997	Oxycon Gamma	2 Maximal GXT: 50W; increased by 50W every 3 min until failure	-0.05 ± 0.16	0.04 ± 0.16	NA	NA
K4	Hausswirth et al., 1994	CPX	Maximal Start @ 100 W for 6 min increased by 25 W every 2 min	Maximal .062 ± .01 .063 ± .01	Maximal 0.074 ± 0.01 0.077 ± 0.011	r^2= 0.953	NA
CosMed $K4b^2$	McLaughlin et al., 2001	DB Method	Rest 50 W 100 W 150 W 200 W 250 W	NA 1.03 vs 0.94 1.63 vs 1.53 2.24 vs 2.14 2.87 vs 2.79 NA	NA NA NA NA significantly lower significantly lower	NA	NA
CosMed $K4b^2$	McNaughton et al., 2005	Morgan EX670	Rest CosMed $K4b^2$ Morgan EX670 Cycling: 150-300W CosMed $K4b^2$ Morgan EX670	 0.41 ± 0.01 0.48 ± 0.02 4.41 L/min 4.51 L/min	 0.42 ± 0.02 0.49 ± 0.02 4.69 L/min 4.94 L/min	NA	NA

(continued)

Table 2. (*continued*)

CosMed K5	Perez-Suarez et al., 2018	Vyntus CPX	rest Cycling 60 W Cycling 130-160 W 13km walk at 5km/h	overestimated 13.4% overestimate 5.8% underestimate 6.6% overestimate 4.8%	NA	NA	NA
Med-Graphics VO2000	Crouter et al., 2006	Parvo-Medics True-One2400	VO2000 ParvoMedics rest 50 W 100 W 150 W 200 W 250 W	0.18 ± 0.15 0.35 ± 0.07 1 ± 0.12 0.96 ± 0.10 1.57 ± .12 1.49 ± 0.09 2.30 ± 0.16 2.15 ± 0.13 3.12 ± 0.32 2.90 ± 0.18 3.88 ± 0.26 3.65 ± 0.18	0.18 ± 0.14 0.32 ± 0.07 0.86 ±0.12 0.87 ± 0.12 1.45 ± 0.12 1.42 ± 0.10 2.19 ± 0.26 2.12 ± 0.22 3.14 ± 0.44 2.94 ± 0.33 4.08 ± 0.5 3.83 ± 0.42	DB: 0.990 Parvos: 0.994 VO2000: 0.989	DB: 0.988 Parvos: 0.991 VO2000 : 0.989
Meta-Max I Meta-Max II	Medbo et al., 2002	DB Method	S1: 2hr cycling power: 2.9W/kg S5: Stepwise manner Power: 30-90% of O_2	S1: no systematic differences in Metamax I vs DB (p = 0.13)	NA	NA	NA

(*continued*)

Table 2. (*continued*)

				S5: 9% higher Metamax I vs DB; 13% higher Metamax II vs DB Metamax II vs Metamax I: systematically higher (p < 0.0001)			
Metamax II	Larsson et al., 2004	DB Method	100 W 200 W 250 W 275 W 300 W 325 W	.03 L /min higher .02 L/min lower .04 L/min higher No difference No difference 0.11 L/min higher	NA	NA	NA
Oxylog	Harrison, Brown, Belyavin, 1982	Standard System	4 experiments conducted: 30-150 W Periods: 3-10min at each work rate RPM: 50	In 2 experiments: Underestimated: 4.4-5.6% Average 4 experiments: Underestimated: 1.5%	NA	NA	NA
CosMed K2	Lothian et al., 1993	Quinton on-line oxygen	1% grade treadmill 4 km/h 12 km/h	5.2% diff 25.8% diff 22.2% diff	NA	0.96--0.99 ICC	NA

(*continued*)

Table 2. (*continued*)

		analyzing system	16 km/h				
K4	Maiolo el al, 2003	Airspec QP9000	rest 8 km/h 10 km/h 12 km/h 14 km/h 16 km/h 18 km/h	0.02 L/min diff 0.11 L/min 0.1 L/min 0.06 L/min 0.18 L/min 0.05 L/min 0.08 L/min	0.01 L/min diff 0.02 L/min 0.09 L/min 0.04 L/min 0.3 L/min 0.11 L/min 0.02 L/min	NA	NA
$K4b^2$	Eisenman et al., 2003	Quark	$K4b^2$ vs Quark				
				80m/min	0.044-0.085L/min	0.0104-0.033L/min	NA
				0% grade; 80m/min	5% difference		
				5% grade	R=0.96	R=0.97	
				10% grade			
CosMed $K4b^2$	Duffield et al., 2004	Morgan Ventilometer and Ametek Applied Electrochemistry	10 min at 10 km/h 3 min at 16 km/h 1 min at 20 km/h	3.32 v 2.86 4.54 v 4.09 3.4 v 2.8	3.09 v 2.77 4.91 v 4.59 2.95 v 2.38	0.85 0.87 0.53	0.81 0.72 0.68
CosMed $K4b^2$	Ross et al., 2019	DB Method	K1 Indoor K2 Outdoor DB	1.8 - 5.5 1.9 - 5.3 1.6 - 5.3	1.4 - 5.5 1.3 - 5.2 1.3 - 5.7	NA	NA
Metamax II	Medbo et al., 2002	DB Method Vmax29	Series 2-4: treadmill stepwise increase S2: Speed: 50-90% of O_2 uptake	S3: (Metamax II and Vmax): no systematic differences	NA	NA	NA

(*continued*)

Table 2. *(continued)*

			S3:Speed: 60-100% of O_2 uptake S4: Speeds: 50-100% O_2 uptake	S4: (Metamax II and DB): Metamax II O2 uptake measured 4% higher than DB ($p= 0.03$)			
K4b[2]	Schrack et al., 2010	Med-graphics D	2.7-4.6mph at 0% grade (self-selected speed)	-0.032 L/min (±0.022)	-0.072 L/min (±0.040)	NA	NA
				ICC	0.97	0.93	NA
				Mean Difference b/w analyzers	-0.033 ± 0.022 L/min	-0.073 ± 0.04 L/min	NA
CosMed K5	White, Deblois, and Barreira 2019	K4 and K5	Treadmill Walk 1.5mph 2.5mph 4.0mph	Range of mean differences: 0.052 ± 0.001	Range of mean differences: 0.047 ±0.08	NA	NA
				Moderate to Strong Relationship: $r = 0.49$-0.80	Moderate to Strong Relationship: $r = 0.51$-0.82		

L/min: Liters per minute; % diff: Percent difference; GXT: Graded exercise test; W: Watts; min: minutes; DB method: Douglas Bag method; RPM: Revolutions per minute; km/h: Kilometers per hour; m/min: Meters per minute; grade: % gradient; r/r²: Pearson's correlation

3.2.1 Validity

Validation studies observed the accuracy of PWMD. Six studies reported no significant differences between portable and criterion devices [11, 31, 33–36]. Conversely, six cycling studies reported significant differences for VO_2 by (0.1–0.3 L/min; 4.4–13% difference) and VCO_2 (0.14–0.3 L/min; 6.9% difference) [10, 13, 32, 37–39]. Four running [27–29, 38] and two walking studies [40, 41] followed similar trends of VO_2 by 0.30–0.63 L/min (5.2–22.2%) and VCO_2 by 0.42–0.62 L/min (3% difference) compared to criterion devices. Generally, as exercise intensity increased, variability between devices correspondingly increased.

Six studies and one review [42] used the Douglas Bag (DB) as a criterion measure, which collected expired air for post-exercise-test analyses through a metabolic cart to acquire air volume (V_E), O_2 and CO_2 gas concentration [10, 25, 29, 43–45]. Five studies compared PWMD versus popular lab-based metabolic carts: 1) CosMed Quark CPET [5, 11] and 2) Parvomedics TrueOne 2400 [9, 10, 46]. Five studies used alternative criteria—mechanical lungs [47, 48] or metabolic simulators [49–51]—to validate the PWMD.

3.2.2 Reliability

Reliability studies used the same device over time to identify measurement consistency. Five studies reported high test-retest reliability of PWMD for VO_2 (ICC: 0.87–0.99) and VCO_2 (ICC: 0.72–0.81) while walking and running at different speeds and for varying gradients [10, 11, 28, 35, 40]. Conversely, one study reported lower test-retest reliability for VO_2 (ICC: 0.53) and VCO_2 (ICC: 0.68) using the CosMed $K4b^2$ while running at high speeds for short distances (1 min at 20 km/h) [27].

4 Conclusion

Portable wearable metabolic devices (PWMD) have been applied to numerous exercise modalities in various environments (see Table 1). Future research should consider expanding PWMD application to team sports and/or other outdoor activities typically not studied using traditional metabolic carts. The application of PWMD has the potential to revolutionize the health/fitness industry by allowing users to measure CRF during real-time outdoor exercise.

Only $\approx$25% of studies showed PWMD were valid compared to criterion measures; therefore, the accuracy of data should be interpreted cautiously (especially at high exercise intensities). However, over 80% of studies showed PWMD were reliable at measuring CRF across time. While PWMD may not replace metabolic carts in research/clinical settings due to ambiguous accuracy, their ability to consistently assess CRF may be more important for users in health/fitness settings. Therefore, future research using PWMD should focus on reliability. Consistent data over time, absent of accuracy, can still be valuable for users tracking CRF progress and performance.

Acknowledgements. The authors wish to thank Jose Mejia, MS for his assistance in translating and revising the Spanish draft version of the abstract of this paper. Figure 1 was illustrated by

Kristen Liang and Eric Gasmin, adapted from 1A) Introducing PNOE VO2Max; 1B) Performance Testing; and 1C) COSMED: The Metabolic Company; referenced from K5_CPET.

References

1. Akalan, C., Kravitz, L., Robergs, R.R.: VO2max: Essentials of the most widely used test in exercise physiology. ACSM's Health Fitness J. **8**(3), 5–9 (2004)
2. Noakes, T.D.: Implications of exercise testing for prediction of athletic performance: a contemporary perspective. Med. Sci. Sports Exerc. **20**(4), 319–330 (1988)
3. Beltz, N.M., Gibson, A.L., Janot, J.M., Kravitz, L., Mermier, C.M., Dalleck, L.C.: Graded exercise testing protocols for the determination of VO_2 max: historical perspectives, progress, and future considerations. J. Sports Med. **2016**, 1–12 (2016). https://doi.org/10.1155/2016/3968393
4. Forman, D.E., Myers, J., Lavie, C.J., Guazzi, M., Celli, B., Arena, R.: Cardiopulmonary exercise testing: relevant but underused. Postgrad. Med. **122**(6), 68–86 (2010)
5. Nieman, D.C., Austin, M.D., Dew, D., Utter, A.C.: Validity of COSMED's quark CPET mixing chamber system in evaluating energy metabolism during aerobic exercise in healthy male adults. Res. Sports Med. **21**, 136–145 (2013)
6. Beijst, C., Schep, G., Breda, E.V., Wijn, P.F., Pul, C.V.: Accuracy and precision of CPET equipment: a comparison of breath-by-breath and mixing chamber systems. J. Med. Eng. Technol. **37**(1), 35–42 (2013)
7. Meyer, T., Georg, T., Becker, C., Kindermann, W.: Reliability of gas exchange measurements from two different spiroergometry systems. Int. J. Sports Med. **22**(08), 593–597 (2001)
8. Roecker, K., Prettin, S., Sorichter, S.: Gas exchange measurements with high temporal resolution: the breath-by-breath approach. Int. J. Sports Med. **26**, S11–S18 (2005). https://doi.org/10.1055/s-2004-830506
9. Bassett, D.R., Jr., et al.: Validity of inspiratory and expiratory methods of measuring gas exchange with a computerized system. J. Appl. Physiol. **91**(1), 218–224 (2001)
10. Crouter, S.E., Antczak, A., Hudak, J.R., DellaValle, D.M., Haas, J.D.: Accuracy and reliability of the ParvoMedics TrueOne 2400 and MedGraphics VO2000 metabolic systems. Eur. J. Appl. Physiol. **98**(2), 139–151 (2006)
11. Eisenmann, J.C., Brisko, N., Shadrick, D., Welsh, S.: Comparative analysis of the Cosmed Quark b (2) and K4b2 gas analysis systems during submaximal exercise. J. Sports Med. Phys. Fitness **43**(2), 150 (2003)
12. Littlewood, R.A., White, M.S., Bell, K.L., Davies, P.S.W., Cleghorn, G.J., Grote, R.: Comparison of the Cosmed K4 b2 and the Deltatrac IITM metabolic cart in measuring resting energy expenditure in adults. Clin. Nutr. **21**(6), 491–497 (2002)
13. McLaughlin, J.E., King, G.A., Howley, E.T., Bassett, D.R., Ainsworth, B.E.: Validation of the COSMED K4 b2 portable metabolic system. Int. J. Sports Med. **22**(4), 280–284 (2001)
14. Vandarakis, D., Salacinski, A.J., Broeder, C.E.: A comparison of COSMED metabolic systems for the determination of resting metabolic rate. Res. Sports Med. **21**(2), 187–194 (2013)
15. Woods, A.L., Garvican-Lewis, L.A., Rice, A.J., Thompson, K.G.: The ventilation-corrected ParvoMedics TrueOne 2400 provides a valid and reliable assessment of resting metabolic rate (RMR) in athletes compared with the Douglas bag method. Int. J. Sport Nutr. Exerc. Metab. **26**(5), 454–463 (2016)
16. Parvataneni, K., Ploeg, L., Olney, S., Brouwer, B.: Kinematic, kinetic and metabolic parameters of treadmill versus overground walking in healthy older adults. Clin. Biomech. **24**(1), 95–100 (2009)

17. Ikegami, Y., Hiruta, S., Ikegami, H., Miyamura, M.: Development of a telemetry system for measuring oxygen uptake during sports activities. Eur. J. Appl. Physiol. **57**(5), 622–626 (1988)
18. Sartor, F., et al.: Estimation of maximal oxygen uptake via submaximal exercise testing in sports, clinical, and home settings. Sports Med. **43**(9), 865–873 (2013)
19. da Cunha, F.A., Farinatti Pde, T., Midgley, A.W.: Methodological and practical application issues in exercise prescription using the heart rate reserve and oxygen uptake reserve methods. J. Sci. Med. Sport **14**, 46–57 (2011)
20. Guimaraes, G., Farinatti, P., Midgley, A., Vasconcellos, F., Vigario, P., Cunha, F.: Relationship between percentages of heart rate reserve and oxygen uptake reserve during cycling and running: a validation study. J. Strength Cond. Res. **33**, 1954–1962 (2019)
21. Capostagno, B., Lambert, M.I., Lamberts, R.P.: Analysis of a submaximal cycle test to monitor adaptations to training: implications for optimizing training prescription. J. Strength Cond. Res. (2019)
22. Garber, C.E., et al.: American college of sports medicine position stand. Quantity and quality of exercise for developing and maintaining, cardiorespiratory, musculoskeletal and neuromotor fitness in apparently healthy adults: guidance for prescribing exercise. Med. Sci. Sports. Exer. **43**(7), 13–34 (2011)
23. Rampichini, S., et al.: Heart rate and pulmonary oxygen uptake response in professional badminton players: comparison between on-court game simulation and laboratory exercise testing. Eur. J. Appl. Physiol. **118**(11), 2339–2347 (2018). https://doi.org/10.1007/s00421-018-3960-6
24. Bekraoui, N., Fargeas-Gluck, M.A., Léger, L.: Oxygen uptake and heart rate response of 6 standardized tennis drills. Appl. Physiol. Nutr. Metab. **37**(5), 982–989 (2012)
25. Kawakami, Y., Nozaki, D., Matsuo, A., Fukunaga, T.: Reliability of measurement of oxygen uptake by a portable telemetric system. Eur. J. Appl. Physiol. **65**(5), 409–414 (1992)
26. Peel, C., Utsey, C.: Oxygen consumption using the K2 telemetry system and a metabolic cart. Med. Sci. Sports Exerc. **25**(3), 396–400 (1993)
27. Duffield, R., Dawson, B., Pinnington, H.C., Wong, P.: Accuracy and reliability of a Cosmed K4b2 portable gas analysis system. J. Sci. Med. Sport **7**(1), 11–22 (2004)
28. Lothian, F., Farrally, M.R., Mahoney, C.: Validity and reliability of the Cosmed K2 to measure oxygen uptake. Can. J. Appl. Physiol. **18**(2), 197–206 (1993)
29. Ross, R., Alduhishy, A., RGonzalez-Haro, C.: Validation of the cosmed K4b2 portable metabolic system during running outdoors. J. Strength Cond. Res. **34**(1), 124–133 (2019)
30. Bigard, A.X., Guezennec, C.Y.: Evaluation of the Cosmed K2 telemetry system during exercise at moderate altitude. Med. Sci. Sports Exerc. **27**(9), 1333–1338 (1995)
31. Larsson, P.U., Wadell, K.M.E., Jakobsson, E.J.I., Burlin, L.U., Henriksson-Larsen, K.B.: Validation of the MetaMax II portable metabolic measurement system. Int. J. Sports Med. **25**(02), 115–123 (2004)
32. Schulz, H., Helle, S., Heck, H.: The validity of the telemetric system CORTEX X1 in the ventilatory and gas exchange measurement during exercise. Int. J. Sports Med. **28**(06), 454–457 (1997)
33. McNaughton, L.R., Sherman, R., Roberts, S., Bentley, D.J.: Portable gas analyzer Cosmed $K4b^2$ compared to a laboratory based mass spectrometer system. J. Sports Med. Phys. Fitness **45**, 315–323 (2005)
34. Doyon, K.H., Perrey, S., Abe, D., Hughson, R.L.: Field testing of VO2peak in cross-country skiers with portable breath-by-breath system. Canadian J. Appl. Physiol. **26**, 1–11 (2001)
35. Hausswirth, C., Bigard, A.X., Le Chevalier, J.M.: The Cosmed K4 telemetry system as an accurate device for oxygen uptake measurements during exercise. Int. J. Sports Med. **18**(6), 449–453 (1997)

36. Maiolo, C.: Physical activity energy expenditure measured using a portable telemetric device in comparison with a mass spectrometer. Br. J. Sports Med. **37**(5), 445–447 (2003). https://doi.org/10.1136/bjsm.37.5.445
37. Harrison, M.H., Brown, G.A., Belyavin, A.J.: The "Oxylog": an evaluation. Ergonomics **25**, 809–820 (1982)
38. Medbø, J.I., Mamen, A., Welde, B., Heimburg, E.V., Stokke, R.: Examination of the Metamax I and II oxygen analysers during exercise studies in the laboratory. Scand. J. Clin. Lab. Invest. **62**(8), 585–598 (2002)
39. Perez-Suarez, I., et al.: Accuracy and precision of the COSMED K5 portable analyser. Front. Physiol. **9**, 1764 (2018)
40. Schrack, J.A., Simonsick, E.M., Ferrucci, L.: Comparison of the Cosmed K4b2 portable metabolic system in measuring steady-state walking energy expenditure. PLoS ONE **5**(2), e9292 (2010). https://doi.org/10.1371/journal.pone.0009292
41. White, L., DeBlois, J., Barreira, T.: Reliability and Validity Analysis of the COSMED K5 Portable Metabolic System. Int. J. Exerc. Sci. Conf. Proc. **9**(7), 115 (2019)
42. Shephard, R.J.: Open-circuit respirometry: a brief historical review of the use of Douglas bags and chemical analyzers. Eur. J. Appl. Physiol. **117**(3), 381–387 (2017). https://doi.org/10.1007/s00421-017-3556-6
43. Ballal, M.A., Macdonald, I.A.: An evaluation of the Oxylog as a portable device with which to measure oxygen consumption. Clin. Phys. Physiol. Meas. **3**, 57–65 (1982)
44. Carter, J., Jeukendrup, A.E.: Validity and Reliability of three commercially available breath-by-breath respiratory systems. Eur. J. Appl. Physiol. **86**, 435–441 (2002)
45. Lucia, A., Fleck, S.J., Gotshall, R.W., Kearney, J.T.: Validity and reliability of the COSMED K2 instrument. Int. J. Sports Med. **14**, 380–386 (1993)
46. Macfarlane, D.J., Wu, H.L.: Inter-unit variability in two ParvoMedics TrueOne 2400 automated metabolic gas analysis systems. Eur. J. Appl. Physiol. **113**(3), 753–762 (2013)
47. Huszczuk, A., Whipp, B.J., Wasserman, K.: A respiratory gas exchange simulator for routine calibration in metabolic studies. Eur. Respir. Soc. **3**(4), 465–468 (1990)
48. Katch, V.L., Sady, S.S., Freedson, P.A.T.T.Y.: Biological variability in maximum aerobic power. Med. Sci. Sports Exerc. **14**(1), 21–25 (1982)
49. Crandall, C.G., Taylor, S.L., Raven, P.B.: Evaluation of the Cosmed K2 portable telemetric oxygen uptake analyzer. Med. Sci. Sports Exerc. **26**(1), 108–111 (1994)
50. Gore, C.J., Catcheside, P.G., French, S.N., Bennett, J.M., Laforgia, J.: Automated VO2max calibrator for open-circuit indirect calorimetry systems. Med. Sci. Sports Exerc. **29**(8), 1095–1103 (1997)
51. Skinner, J.S., et al.: Reproducibility of maximal exercise test data in the HERITAGE family study. Med. Sci. Sports Exerc. **31**(11), 1623–1628 (1999)

My Proprioception, a Smartphone Application to Evaluate Proprioception in Sport: Proprioceptive Profiles of Female Players of Basketball Balearic League of Mallorca Island

Andreu Sastre-Munar[1], Juan Carlos Fernández-Domínguez[2], and Natalia Romero-Franco[2](✉)

[1] Sport Technification Center "Príncipes de España", 07009 Palma de Mallorca, Spain
[2] Nursing and Physiotherapy Department, University of the Balearic Islands, 07122 Palma de Mallorca, Spain
natalia.romero@uib.es

Abstract. Proprioception, as parameter of motor control, is important for Women's Basketball due to the high prevalence of injuries. To date, proprioceptive evaluation throughout joint position sense (JPS) had required sophisticated instruments or time-consuming tasks. Therefore, the use of JPS as part of sport evaluation had been limited. The new technologic advances have enabled the development of smartphone applications to obtain JPS in a valid, reliable and rapid way. We aimed to describe the proprioceptive profiles for knee joint position sense of female basketball players of Mallorca island. For this, 91 healthy female basketball players (18 to 30 years old) were evaluated during the basketball preseason. All participants performed a JPS test of knee-joint in closed kinetic chain. Also, anthropometric data and sport-related information were collected. Data showed 4.0 ± 3.4 y 4.1 ± 3.0 degrees as proprioceptive error for right and left knee, respectively, without significant differences according to dominance or player position ($p > 0.05$). No significant correlations were obtained between JPS and age or anthropometric variables. As a conclusion, the proprioceptive profiles of female players from Basketball Balearic League of Mallorca island are about 4.0 degrees of proprioceptive error, without being influenced by dominance, player position, age or anthropometric variables.

Keywords: Women's Basketball · Proprioception · Knee

1 Introduction

Sport injuries due to lack of motor control is a frequent event in basketball, with even a higher incidence if we focus on women's basketball [1]. In this term, epidemiological studies report that about 40% of female players suffer of ankle injuries, and also that most knee injuries are caused by non-contact mechanisms [1, 2].

To decrease incidence of injuries and to increase athlete's performance, the researchers advise improving motor control [3] because deficiency of the latter seems to increase risk of injuries and decrease performance [4].

B. Sañudo Corrales and J. García-Fernández (Eds.): Tapasconference 2020, LNBE, pp. 26–32, 2022.
https://doi.org/10.1007/978-3-030-92897-1_3

As a key parameter of motor control, we must take into account the proprioception, which can be evaluated through joint position sense (JPS) [5]. However, until now it had not been included as a part of sports evaluation due to the great difficulty in terms of accessibility to instruments [6] or required time-consuming tasks [7].

The new technologic advances have enabled the development of smartphone applications to obtain JPS in a valid and reliable way, at the same time that allow to assess the range of movement (ROM), showing immediate results [8]. In this sense, the smartphone application *MyProprioception* has been specifically designed to perform instant proprioceptive evaluation enabling to radically decrease evaluation time and making its measurement possible on the field, in a valid and reliable way [8].

This way, this new tool could supply a large extent of athlete's evaluation and it would help to obtain proprioceptive data as guidance during female basketball evaluation. Therefore, this study was aimed to describe joint position sense profiles of knee in women's basketball players from basketball Balearic league of Mallorca island.

2 Methodology

2.1 Design

An observational study was designed to evaluate knee proprioception of female basketball players from Mallorca island. This study took place during September, 2019, which referred to the preseason of the female basketball players. Also, evaluations were carried out in the sport facilities of the basketball teams, always during the afternoon to avoid differences due to the time of day. Therefore, three weeks were needed to collect all data due to the high number of participants.

2.2 Participants

Twelve female basketball teams from Basketball Balearic League voluntary participated in this study. As eligibility criteria, participants should have at least 2 years of experience as basketball players in this level of competition, have between 18 and 30 years old, be free of injuries during at least 6 months prior to the beginning of study, and not have had any surgical intervention due to lower extremity injuries. 91 female basketball players met the inclusion criteria and accepted to participate.

Before participation, all basketball players were informed about the study and signed an informed consent form according to the Declaration of Helsinki. The ethics committee of the local university approved this study.

2.3 Procedures

Anthropometric information was assessed in all the participants by evaluating body mass and height using a ± 100-g precision digital weight scale (Tefal, France) and a t201-t4 adult height scale (Asimed, Spain), respectively. Also, all players provided data on a questionnaire about their age, experience as basketball player, and leg dominance (leg to kick a ball). During this same session, all players were asked to perform a knee proprioception test.

Knee Proprioception. All players were asked to perform a JPS test in passive–active modality and closed-kinetic chain during knee flexo-extension, in both left and right knee joints. Order of tests (to evaluate right or left knee joints) was always randomized with a coin. This test modality and knee joint were selected due to the intermediate position of the knee in lower limb and the biomechanical requirement from the whole lower extremity in the closed-kinetic chain movement. The JPS procedure was performed by using the MyProprioception app previously installed on an iPhone X (iOS 11.3, Apple Inc., USA), which has been recently validated [8]. The functioning of this smartphone application is based on photo-analysis for the joint angle assessment. For the JPS test, procedure was similar to previous studies [9]. Before beginning the test, the iPhone X was positioned in vertical format on a tripod located 1 m distance away from the participant and 1 m height from the floor. Once iPhone was ready, participants were assessed from standing as initial position, with visual inputs blocked with a mask, and were guided to a target position (approximately 50° of knee flexion). This joint angle was maintained during 5 s to allow participants to recognize it before returning to initial position. After the "reposition" order, participants had 3 trials to reproduce the target angle previously memorized, as close as possible. While participants performed test, photos were taken with the iPhone application for the target position and the 3 trials positions. In this way, just after the test, photo-analysis was completed by using the iPhone application to obtain the knee angle for every position. With this information, the iPhone application automatically reported the absolute angular error (AAE). This procedure was completed for both left (AAE_L) and right (AAE_R) knee joints. This error is the most reported among scientific literature because it informs about the absolute difference between target position and the mean of the three trials, without considering the direction of the difference [10].

Before the proprioception test, all players should perform a 15-min warm-up consisting on 5-min jogging, 5-min dynamic stretching, and 5-min racing exercises. After warming-up, participants completed the familiarization process by performing a knee JPS test.

2.4 Statistical Analysis

Mean and standard deviations for continuous variables, as well as frequencies for categorical one were obtained. One-way ANOVA was used to evaluate differences in AAE according to leg dominance (right or left) and also according to playing position (point guard, shooting guard, power forward, center) The Pearson r was used to evaluate correlations between age and AAE, and between body mass index (BMI) and AAE. The r Pearson values interpretation was as follows: low (< 0.5), moderate (0.51–0.70), high (0.71–0.90), very high (>0.91) [11]. The SPSS software v.21 (Chicago, USA) was used. The level of significance was established at $p < 0.05$.

3 Results

Table 1 shows descriptive data from anthropometric and proprioceptive variables of the female basketball players.

Table 1. Anthropometric and proprioceptive values

		Players (n = 91)
		Mean (SD)
Age (years)		21.1 (3.4)
Weight (kg)		67.7 (13.1)
Height (m)		1.69 (0.07)
BMI (kg/m^2)		23.0 (5.2)
Playing position (%)	Point guard	29.7
	Shooting guard	38.5
	Power forward	8.8
	Centre	23.0
	Right	85.7
Dominance (%)	Left	14.3
AAE_R (degrees)		4.0 (3.4)
AAE_L (degrees)		4.1 (3.1)

AAE_L = absolute angular error in left knee; AAE_R = absolute angular error in right knee; BMI = body mass index; SD = standard deviation; SD = standard deviation

As Table 2 shows, no statistically significant differences were found in JPS according to the dominance (AAE_R: F = 0.90, p = 0.765; AAE_L: F = 0.954, p = 0.331) or playing position (AAE_R: F = 1.109, p = 0.349; AAE_L: F = 1.334, p = 0.268). Also, no statistically significant correlations were shown between age and AAE_R (r = –0.003, p = 0.977) or AAE_L (r = 0.056, p = 0.564), or between BMI and AAE_R (r = 0.061, p = 0.531) or AAE_L (r = –0.057, p = 0.560).

Table 2. Proprioceptive differences according to dominance and play position

		AAE_R (degrees) *Mean (SD)*	AAE_L (degrees) *Mean (SD)*
Dominance	Right (n = 78)	4.2 (3.5)	3.2 (2.4)
	Left (n = 13)	4.1 (3.2)	4.3 (2.9)
	Point guard (n = 27)	4.8 (4.5)	4.6 (2.8)
	Shooting guard (n = 35)	3.5 (2.8)	4.0 (3.5)
	Power forward (n = 8)	2.5 (1.2)	2.4 (2.7)
Playing position	Centre (n = 21)	4.3 (2.9)	4.3 (3.2)

AAE_L = absolute angular error in left knee; AAE_R = absolute angular error in right knee; SD = standard deviation

4 Discussion

The main findings of this study show that female basketball players had error proprioceptive values less than 5° for knee joint, without differences according to anthropometric characteristics, dominance or playing position.

The findings are similar to proprioceptive values reported in previous studies that reported approximately 4° as normal values for absolute angular error [12]. It is necessary to consider that authors who reported reference values of proprioception had always considered healthy and sedentary population. In this sense, the level and type of physical exercise is a factor that clearly influences joint position sense [12, 13]. In this line, our results show worse proprioceptive error values than those found in soccer players or volleyball players [13, 14]. Besides, our study assessed female amateur players, while most of the previous studies considered semi-professional or professional players. This fact would support previous studies that suggested competition level as an important condition to achieve better proprioceptive values, despite belonging to the same sport [15].

With regard to the evaluation process, other parameters could have influenced results such as existence or lack of warm-up, the passive or active modality of positioning, number of attempts used or participant's positioning during the test [16]. According to previous studies, warm-up could improve proprioceptive values [17] and these are more reliable when the modality is active and in closed kinetic chain [18].

Regarding to the absence of correlations with age, we should highlight that participants considered in our study had a relative short age range (between 18 and 30 years old). In fact, previous studies that evaluated differences according to age, considered ranges from 20 to 45 years old as the same range of age [12]. In the same line, we did not find either differences according to anthropometric variables because our players had similar values of BMI. Previous studies have shown proprioceptive deficit only when BMI was very high (>25 kg/m^2) [19].

For leg dominance, studies to date have not obtained differences either, thus our results would follow the same line [20]. Moreover, playing position did not show any influence on knee proprioception of female players in our study. However, we observed that power forward position players showed a tendency to a better joint position sense. Nevertheless, bearing in mind that number of players who occupied that position was very low in our study, additional studies are needed to confirm this trend.

This study had some limitations. As major limitation, standard deviation values were very high and this could difficult the clinical interpretation. These results were expected due to the high variability observed in the proprioceptive evaluation between the participants, as previous studies obtained [12]. Additional studies are necessary to evaluate proprioceptive error in female basketball players with previous injuries or to evaluate risk of injuries later on.

As practical applications, sports and health professionals could consider proprioceptive measurement as part of the routine evaluations of female basketball players thanks to the existence of tools that enables quick evaluation in field conditions. These values could be used to monitor female basketball players and design appropriate motor control strategies, pending future studies that demonstrate the capability of the proprioceptive errors to predict injuries.

5 Conclusion

The proprioceptive profiles of female players from basketball Balearic league of Mallorca island are about 4.0 degrees of proprioceptive error, without being influenced by dominance, player position, age or anthropometric variables.

References

1. Tummala, S.V., Hartigan, D.E., Makovicka, J.L., Patel, K.A., Chhabra, A.: 10-year epidemiology of ankle injuries in men's and women's collegiate basketball. Orthopaed. J. Sports Med. **6**(11), 232596711880540 (2018). https://doi.org/10.1177/2325967118805400
2. Pasanen, K., et al.: High ankle injury rate in adolescent basketball: A 3-year prospective follow-up study. Scand J Med Sci Sports **27**(6), 643–649 (2017)
3. Benis, R., Bonato, M., La Torre, A.: Elite female basketball players' body-weight neuromuscular training and performance on the y-balance test. J. Athl. Train. **51**(9), 688–695 (2016)
4. Chimera, N.J., Smith, C.A., Warren, M.: Injury history, sex, and performance on the functional movement screen and Y balance test. J. Athl. Train. **50**(5), 475–485 (2015)
5. Takasaki, H., Okubo, Y., Okuyama, S.: The effect of proprioceptive neuromuscular facilitation on joint position sense: a systematic review. J. Sport Rehabil. **29**(4), 488–497 (2019)
6. Hillier, S., Immink, M., Thewlis, D.: Assessing proprioception: a systematic review of possibilities. Neurorehabil. Neural Repair. **29**(10), 933–949 (2015)
7. Naseri, N., Pourkazemi, F.: Difference in knee joint position sense in athletes with and without patellofemoral pain syndrome. Knee Surg. Sport Traumatol. Arthrosc. **20**(10), 2071–2076 (2012)
8. Romero-Franco, N., Jiménez-Reyes, P., González-Hernández, J.M., Fernández-Domínguez, J.C.: Assessing the concurrent validity and reliability of an iPhone application for the measurement of range of motion and joint position sense in knee and ankle joints of young adults. Phys. Ther. Sport. **44**, 136–142 (2020)
9. Romero-Franco, N., Montaño-Munuera, J.A., Jiménez-Reyes, P.: Validity and reliability of a digital inclinometer to assess knee joint-position sense in a closed kinetic chain. J. Sport Rehabil. **26**(1), 1–14 (2017). https://doi.org/10.1123/jsr.2015-0138
10. Röijezon, U., Clark, N.C., Treleaven, J.: Proprioception in musculoskeletal rehabilitation. Part 1: basic science and principles of assessment and clinical interventions. Manual Ther. **20**(3), 368–377 (2015). https://doi.org/10.1016/j.math.2015.01.008
11. Hinkle, D.E., Wiersma, W., Jurs, S.G.: Applied statistics for the behavioral sciences [Internet]. Houghton Mifflin (2003). (Applied Statistics for the Behavioral Sciences). https://books.google.es/books?id=7tntAAAAMAAJ
12. Relph, N., Herrington, L.: The effects of knee direction, physical activity and age on knee joint position sense. Knee **23**(3), 393–398 (2016)
13. Ribeiro, F., Oliveira, J.: Effect of physical exercise and age on knee joint position sense. Arch. Gerontol. Geriatr. **51**(1), 64–67 (2010)
14. Salgado, E., Ribeiro, F., Oliveira, J.: Joint-position sense is altered by football pre-participation warm-up exercise and match induced fatigue. Knee **22**(3), 243–248 (2015)
15. Waddington, G., Han, J., Adams, R., Anson, J.: Measures of proprioception predict success in elite athletes. J. Sci. Med. Sport. **16**, e19 (2013)
16. Daneshjoo, A., Mokhtar, A.H., Rahnama, N., Yusof, A.: The effects of comprehensive warm-up programs on proprioception static and dynamic balance on male soccer players. PLoS One. **7**(12), e51568 (2012)

17. Subasi, S.S., Gelecek, N., Aksakoglu, G.: Effects of different warm-up periods on knee proprioception and balance in healthy young individuals. J Sport Rehabil. **17**(2), 186–205 (2008)
18. Bennell, K., Wee, E., Crossley, K., Stillman, B., Hodges, P.: Effects of experimentally-induced anterior knee pain on knee joint position sense in healthy individuals. J. Orthop. Res. **23**(1), 46–53 (2005)
19. Numanoğlu, E.A., Can, F., Erden, Z.: Do Body mass index and body fat ratio have an effect on proprioception? Orthop. J. Sport Med. **2**(11_suppl3), 1–3 (2014). https://doi.org/10.1177/2325967114S00151
20. Cug, M., Wikstrom, E.A., Golshaei, B., Kirazci, S.: The effects of sex, limb dominance, and soccer participation on knee proprioception and dynamic postural control. J .Sport Rehabil. **25**(1), 31–39 (2016)

The Impact of Smartphone Use on Body Composition, Physical Fitness, Quality of Life and Selective Attention on Office Workers. A Pilot Study

Laura Velasco-Llorente(✉) and Borja Sañudo

University of Seville, 41013 Seville, Spain
lauvelllo@alum.us.es, bsancor@us.es

Abstract. Office workers spend a lot of time in front of screens at the workplace and at home. Nowadays the smartphone is the most used device, and it is not clear how it can affect our health and wellbeing. Consequently, the objective of this research was to analyse the relationship between smartphone use and body composition (BC), physical fitness (PF), quality of life (QoL) and selective attention (SA) in office workers. A cross-sectional pilot study was carried out with 30 office workers (16 men and 14 women) who were between 27 and 59 years old. We collected their BC (height, weight, body mass index, waist-hip ratio -WHR- and fat mass percentage), PF (countermovement jump –CMJ-, handgrip strength and cardiovascular endurance performing the 6-min walk test), QoL (SF-36), SA (Stroop Test) and smartphone use. As a result, smartphone use was significantly correlated with QoL dimensions, pain ($r^2 = 0.219$, $p = 0.04$) and physical component ($r^2 = 0.153$, $p = 0.03$). Smartphone use was also correlated with the congruent score ($r^2 = 0.143$, $p = 0.01$) and the "Stroop Effect" ($r^2 = 0.195$, $p < 0.01$). The results of this pilot study suggest that smartphone use could correlate with QoL dimensions and SA. It will be necessary to continue and expand the study to acquire more knowledge about how these devices affect health and wellbeing.

Keywords: Smartphone · Office worker · Quality of life · Physical fitness · Cognitive ability

1 Introduction

Nowadays, technology is considered an essential part of our daily lives. The smartphone is the most used device whose users spend an average of 800 h per year surfing on the Internet [1]. Data from this report showed that Spain is one of the countries with the highest number of smartphone users. Moreover, they used them an average time of more than 3 h per day. Despite the importance of smartphones in our lives, it is not clear how it can affect our health and wellbeing. Some studies with children have suggested that there is a relationship between its use and insomnia [2] or mental distress [3]. In young adults, some authors have described headache, neck and shoulder pain [4, 5]

B. Sañudo Corrales and J. García-Fernández (Eds.): Tapasconference 2020, LNBE, pp. 33–42, 2022.
https://doi.org/10.1007/978-3-030-92897-1_4

associated with this use. Furthermore, Chrisman, Chow, Daniel, Wu, and Zhao [6] also found an association with sitting time, less physical activity, and an unhealthy body composition. Although there are recent studies about this topic, several of them reported inconsistent conclusions because they are limited due to the lack of objective measuring instruments. Also, there are few studies with samples of adults. Consequently, to increase the knowledge on the impact of the use of smartphones on adults, this study aimed at assessing a sample of office workers because their job is very related to the screen usage and they are sitting for a long time in the workplace [7]. Therefore, the impact of an excessive use of smartphones could further worsen their health.

2 Objective

The aim of this study is to determine the association between smartphone use and Body Composition (BC), Physical Fitness (PF), Quality of Life (QoL) and Selective Attention (SA) in office workers.

3 Methods

3.1 Participants

This cross-sectional study was made with a convenience sample of 30 office workers (16 males and 14 females) belonging to a municipal company of Seville (Spain). It was conducted between March and May 2019. All participants provided written consent and agreed to the data protection declaration. To be included, the worker had to be between 18 and 65 years old. Moreover, they had to have been in the current position for at least six months and they had to perform administrative tasks for a high proportion of their working hours. Participants were excluded if they were unable to maintain their routine while the data was being collected.

3.2 Procedures

The study was carried out during different weeks of March and May 2019. The participants were divided into groups and the data of each group was collected in a week. The researchers were at the office on the first and seventh day of each week of the study. The study and procedures were explained on the first day. If participants agreed with the protocol, they provided written consent and the data protection declaration. Subsequently, participants had to fill in a demographic questionnaire which included age, gender, and some questions about their occupation (experience, timetable, tasks, etc.). The same day, they also completed the Stroop Color and Word Test. Afterwards, participants were instructed to download and adjust the App which was going to collect the smartphone use: "Your Hour" for Android and "Screen Time" for iOS. A week after, the researchers went back to the office and completed the study. BC was determined by bioimpedance and anthropometric measurements. Secondly, participants completed a series of PF tests including hand-grip strength, counter-movement jumps (CMJ), and the 6-min walk test. Also, QoL was assessed (SF-36 short form) using an online platform. Finally, researchers verified "Your Hour" or "Screen Time" Apps to obtain the amount of unlocks and minutes spent on the smartphone each day of this week.

3.3 Outcome Measures

First, demographic information (sex, age, civil status etc.) was collected and information related to occupation (tasks, average number of hours per day, category etc.) was also recorded. The smartphone use was objectively assessed. In addition, dependent variables were BC, PF, QoL and SA. BC, PF and SA were objectively measured by bioimpedance, fitness tests and Stroop Test respectively. The short form SF-36 was used to know self-reported QoL.

Smartphone Use. The smartphone screen-state was objectively captured for all users through the screen-state sensor that recorded the number of times that the screen was turned on and the total time use. the smartphone use accumulated each day during the week was recorded by means of the apps "Your Hour" for android (Mind-e-FY Solutions, Madhya Pradesh, India) or "Screen Time" for IOs (a built-in application on the iPhone). Subsequently, an average number of minutes and unlocks per day was obtained.

Anthropometric Measures and Body Composition. Height and weight of participants were used to calculate body mass index (BMI). In addition, their waist and hip measurements were taken to assess waist–hip ratio (WHR). Moreover, percentages of fat mass and kilograms of muscle mass were measured by bioimpedance (bodystat 1500).

Physical Fitness. Participants did three tests to value their strength and cardiorespiratory fitness. First, hand grip strength was assessed using a manual dynamometer (Camry EH101) three times with both hands. secondly, after some practice, participants did three CMJ [8] on the chronojump contact platform [9]. The maximum values of each test were chosen for the statistical analysis. finally, they did the 6-min walking test [10, 11].

Quality of Life. The Spanish version of the SF-36 health survey was used to value self-reported QoL. The form generates 8 subscales and 2 scores: a physical component and a mental component. The subscales are physical functioning, role limitations due to physical problems, pain, general health perception, vitality, social functioning, role limitations due to emotional problems, and mental health [12].

Selective Attention. Participants did the Stroop Test on a computer. For online data collection, the software PsyToolkit was used [13, 14]. The test reports the time spent on congruent answers (e.g., the color of the word "blue" printed in blue ink color), incongruent answers (e.g., the color of the word "blue" printed in yellow ink color), and incongruent menus congruent scores (Stroop Effect). The Stroop test was used to value their ability to inhibit cognitive interference [15, 16].

3.4 Statistical Analysis

Data analysis was performed using IBM SPSS Statistics for Windows (Version 23.0., IBM Corp, Armonk, NY, USA). descriptive analyses were presented as means and standard deviations for continuous ordinal variables, and as frequencies and percentages for categorical variables. spearman's correlation coefficients were calculated to reflect the distribution and the relationships between variables. Alpha was set at $P < 0.05$ for tests.

4 Results

The characteristics of the subjects are showed in Table 1. There were 30 participants (16 males and 14 females). The average age was 46.9 ± 8.59 years (males: 47.8 ± 6.8 years; females: 45.86 ± 10.4). Although, more than 75% of the sample were over 40 years old. The participants were office workers; particularly, 66% of them were technical officers. The most of them had a permanent contract (93.3%) and their work shifts were continuous (93.3%) with an average of more than 31 working hours per week (93.3%). In addition, 90% of the participants had been in that workplace for more than 6 years.

Secondly, women and men showed a different use of smartphones. While the average number of unlocks per day were 45.1 ± 24.7 for men, the average for women was 61.6 ± 32.7. In fact, 62.5% of men completed less than 50 unlocks per day while more than 57.1% of women completed between 50 and 100 unlocks per day. In the same way, the time spent on smartphone was less for men (112.1 min) than for women (130.2 min). Accordingly, more than 40% of women had spent more than 150 min but only 18.8% of men had spent those minutes on the smartphone.

Data about BC is presented in Table 2. According to BMI, 36.7% of the participants presented normal weight but 23.3% were overweight. In this case, 31.3% of men had a BMI over 30.0 while it happened to 14.2% of women. Also, 43.8% of men and 14.3% of women had a WHR over recommendations. The PF values showed differences between women and men in all the variables (Table 3).

The QoL (SF-36) indicated that participants had good perceptions about their lives, their mean scores were 88.1 (men) and 80.2 (women). Men rated better in general but 46% of them reported that they had suffered pain in those weeks while only 18% of women reported that. Moreover, women presented more limitations due to their physical function (89.2) and their emotions (80.9) than men (100.0 and 94.8 respectively). Stroop Test scores showed a mean value of 931 ± 161 ms and 964 ± 166 ms of time spent on congruent slides and incongruent slides, respectively. The subtraction between these two values resulted in a Stroop effect of 30.4 ± 78.7 ms.

The correlational analysis presented some significant relationships. First, the amount of unlocks were significantly related to the age of the users ($r^2 = 0.128$, $p = 0.02$). There were a higher number of unlocks per day amongst younger participants in comparison to the older participants. On the other hand, BC and PF variables did not have any significant correlation with the smartphone use.

Regarding the QoL, there were significant correlations between the number of unlocks and the pain score ($r^2 = 0.219$, $p = 0.04$) of the SF-36. Additionally, similar results were presented between physical component and time usage ($r^2 = 0.153$, $p = 0.03$). That relationship would suggest that users who unlocked less the smartphone would have a better perception about pain and their physical ability. The mental component was not significantly correlated with the smartphone use; only the negative correlation between unlocks and vitality could be underlined ($r^2 = 0.110$, $p = 0.08$).

Table 1. Characteristics of the participants in the study. Data are presented as frequencies (%)

	All (n = 30)	Males (n = 16)	Females (n = 14)
Age			
20 – 29 years	2 (6.7%)	-	2 (14.3%)
30 – 39 years	4 (13.3%)	2 (12.5%)	2 (14.3%)
40 – 49 years	8 (26.7%)	5 (31.3%)	3 (21.4%)
50 – 59 years	16 (53.3%)	9 (56.3%)	7 (50.0%)
Average (±SD)	46.9 (8.5)	47.8 (6.8)	45.8 (10.4)
Occupation			
Scholarship	2 (6.7%)	-	2 (14.3%)
Technical officer	20 (66.7%)	9 (56.3%)	11 (78.6%)
Middle level	5 (16.7%)	4 (25.0%)	1 (7.1%)
Head of department	3 (10.0%)	3 (18.8%)	-
	All (n = 30)	**Males (n = 16)**	**Females (n = 14)**
Contract			
Apprenticeship	1 (3.3%)	-	1 (7.1%)
Temporary contract	1 (3.3%)	-	1 (7.1%)
Permanent contract	28 (93.3%)	16 (100.0%)	12 (85.7%)
Hours of working per week			
21 – 25 hours	2 (6.7%)	-	2 (14.3%)
26 – 30 hours	-	-	-
31 – 35 hours	19 (63.3%)	10 (62.5%)	9 (64.3%)
< 35 hours	9 (30.0%)	6 (37.5%)	3 (21.4%)
Shift			
Continuous work	28 (93.3%)	15 (93.8%)	13 (92.9%)
Shift work	2 (6.7%)	1 (6.3%)	1 (7.1%)
Years in current position			
< 1 year	2 (6.7%)	-	2 (14.3%)
1 – 3 years	-	-	-
3 – 6 years	1 (3.3%)	1 (6.3%)	-
> 6 years	27 (90.0%)	15 (93.8%)	12 (85.7%)

Finally, the evaluation of SA had some significant correlation with the smartphone use which are summarized in Table 4. The time spent on congruent answers presented a negative significant relationship with the average of unlocks per day ($r^2 = 0.143$, $p = 0.01$). This would mean that participants who unlocked the smartphone more times were faster at answering congruent slides than those who had reported less smartphone use. A more complex variable is the "Stroop effect", it was significantly correlated

Table 2. Anthropometric and body composition measures. The data are presented as means (±SD) and frequencies (%)

	All (n = 30)	Males (n = 16)	Females (n = 14)
Heigh (cm)	169.2 (9.8)	174.8 (8.1)	162.8 (7.6)
Weigh (kg)	77.9 (18.2)	85.6 (19.0)	69.0 (12.8)
BMI			
Underweight	-	-	-
Normal weight	11 (36.7%)	5 (31.3%)	6 (42.9%)
Pre-obesity	12 (40%)	6 (37.5%)	6 (42.9%)
Obesity	7 (23.3%)	5 (31.3%)	2 (14.2%)
Overage ($kg \cdot m^{-2}$)	27.0 (4.85)	27.8 (5.1)	26.0 (4.5)
Waist (cm)	91.0 (14.3)	97.3 (13.7)	83.9 (11.6)
Hip (cm)	104.5 (9.9)	106.0 (10.7)	102.9 (9.17)
WHR (waist-to-hip ratio)			
Negative value	9 (30.0%)	7 (43.8%)	2 (14.3%)
Positive value	21 (70.0%)	9 (56.3%)	12 (85.7%)
Overage	0.86 (0.07)	0.91 (0.05)	0.81 (0.06)
Fat mass (%)	30.9 (8.7)	25.3 (6.3)	37.7 (6.0)
Muscle mass (kg)	54.0 (12.6)	62.7 (9.0)	43.2 (6.5)

Table 3. Physical fitness values. The data are presented as means (±SD).

	All (n = 30)	Males (n = 16)	Females (n = 14)
Handgrip strength (kg)			
Right	34.9 (11.2)	42.8 (8.7)	26.0 (5.4)
Left	33.7 (11.6)	42.5 (8.3)	23.8 (4.6)
CMJ (countermovement jump) (cm)	18.1 (5.2)	21.7 (4.4)	14.0 (2.4)
6' Walking Test (m)	636.8 (53.7)	658.8 (60.2)	611.6 (31.24)

with the smartphone unlocks ($r^2 = 0.130$, $p = 0.02$) and time usage ($r^2 = 0.195$, $p < 0.01$). Knowing that the "Stroop effect" is the incongruent score minus congruent score, those findings would suggest that those who has reported to have used less the smartphone obtained a more negative "Stroop effect" which means that they could inhibit the meanings better than high smartphone users.

Table 4. Spearman's correlation between selective attention and smartphone use ($n = 29$)

Selective attention	Amount of unlocks	Usage time
Congruent score	-0.478^{a}	-0.356
Incongruent score	-0.259	-0.151
Stroop effect score	-0.433^{a}	-0.487^{b}

[a] Correlation significant at 0.05 level
[b] Correlation significant at 0.01 level

5 Discussion

In the present pilot study, we analysed objectively how smartphone use was associated with BC, PF, QoL and SA in a sample of office workers. To date, the evidence is scarce regarding the relationships between these variables. Moreover, the results in previous studies have usually been obtained by subjective measures in samples of children or teenagers [4, 17, 18]. Therefore, the methods in the present study have been able to find the relationship between the smartphone use and QoL and SA in office workers.

Regarding the findings on BC and PF, the results of this research have been inconclusive. Similarly, Biddle et al. [19] did not find a clear relationship between screen use and obesity risk or adiposity in adults. They suggested that the screen use might be a risk factor if it was a sedentary activity (e.g., watching TV). Despite that, other studies have suggested that the smartphone use is related to sedentary behaviour [20, 21]. This relationship seems clearer in young people; recently, Fang, Mu, Liu, and He, advised that "the screen time for more than two hours/day may be one of the risk factors for being overweight or obesity in children and adolescents" [22] and they include the smartphone use. The current data cannot clarify these relationships, so further research is warranted on this topic to know more about the impact of smartphones on BC and physical fitness on adults.

On the other hand, our findings can report some information about the smartphone use and the QoL. The results suggest that those participants with low self-reported QoL are more likely to increase the use of their smartphones. Specifically, it was associated with more pain and worse perception of physical function. In agreement with this data some authors have valued a QoL deficiency in teenagers due to an overuse of screens [2, 5]. In addition, there are studies whose researchers reported more frequent backaches [4] and headaches [23] in people who use the smartphone for more time. Despite not findings of this study had significant relationships with the mental component of QoL, other researchers have suggested that high levels of smartphone use may also be associated with mental distress [3]. Future studies should amplify the research to assess this relationship.

Finally, in the current study, we could observe that faster responses on congruent slides were correlated with a higher smartphone use. However, the Stroop effect score was worse in those with a higher smartphone use. In other words, those who reported a higher use of smartphone would be faster at answering congruent slides, but they would inhibit the information badly. Similarly, Abramson et al. [24] reported that participants who spent more time on the mobile phone needed longer time to complete the Stroop Test.

Although, the knowledge about this topic is deficient because the investigations about smartphone use are contradictory [25]. According to our results, we agree with Wilmer et al. [25] who explained that it is necessary more researches about the smartphone and its impact on cognitive ability because these devices have multiple and complex functions.

To sum up, our findings suggest that a high exposure to smartphones in office workers might be correlated with a lower level of QoL and worse SA. Consequently, it could be hypothesized that a reduction in smartphone use could improve those factors. However, as a pilot cross-sectional study, the results of this study must be considered with caution. The main limitations of this study were its design because it was not experimental, and we could not make causal inferences. Also, the sample is not representative, and the results cannot be generalized. Moreover, the smartphone use measurement must be detailed better to avoid taking not real time use (e.g., collecting as "time use" when the smartphone is on, but it is not being used by the user).

6 Conclusions

Despite the limitations, it can be concluded that there is an association between objectively measured smartphone use and QoL and SA in a sample of office workers. The participants who spent more time using the smartphone were more likely to report worse QoL perception (pain and physical function dimensions) and, although they were faster at answering congruent slides, they inhibited the information badly. With that findings, the researchers have tried to introduce new ideas and give rise to reflections about the use of technologies and how they can impact on the health, behaviours, and cognitive ability of users.

References

1. Rivero, F.: Informe di trendia: Mobile en España y en el Mundo 2019. Inf. técnico, Ditrendia. (2019)
2. Tamura, H., Nishida, T., Tsuji, A., Sakakibara, H.: Association between excessive use of mobile phone and insomnia and depression among Japanese adolescents. Int. J. Environ. Res. Public Health. **14**, 1–11 (2017). https://doi.org/10.3390/ijerph14070701
3. Liu, Y., et al.: Problematic smartphone use and its relation with anxiety and depression: a survey in a large sample of undergraduates. Chin. J. Psychol. **50**, 270–275 (2017). https://doi.org/10.3760/cma.j.issn.1006-7884.2017.04.008
4. Al-Hadidi, F., et al.: Association between mobile phone use and neck pain in university students: a cross-sectional study using numeric rating scale for evaluation of neck pain. PLoS One. **14**, 1–10 (2019). https://doi.org/10.1371/journal.pone.0217231
5. Durusoy, R., Hassoy, H., Özkurt, A., Karababa, A.O.: Mobile phone use, school electromagnetic field levels and related symptoms: a cross-sectional survey among 2150 high school students in Izmir. Environ. Heal. **16**, 1–14 (2017). https://doi.org/10.1186/s12940-017-0257-x
6. Chrisman, M., Chow, W.-H., Daniel, C.R., Wu, X., Zhao, H.: Mobile phone use and its association with sitting time and meeting physical activity recommendations in a Mexican American cohort. JMIR mHealth uHealth. **4**, 1–11 (2016). https://doi.org/10.2196/mhealth.4926

7. Macdonald, B., Janssen, X., Kirk, A., Patience, M., Gibson, A.M.: An integrative, systematic review exploring the research, effectiveness, adoption, implementation, and maintenance of interventions to reduce sedentary behaviour in office workers. Int. J. Environ. Res. Public Health. **15**, 1–29 (2018). https://doi.org/10.3390/ijerph15122876
8. Bosco, C., Luhtanen, P., Komi, P.V.: A simple method for measurement of mechanical power in jumping. Eur. J. Appl. Physiol. Occup. Physiol. **50**, 273–282 (1983)
9. De Blas, X., Padullés, J.M., Del Amo, J.L.L., Guerra-Balic, M.: Creación y validación de Chronojump-Boscosystem: un instrumento libre para la medición de saltos verticales. RICYDE Rev. Int. Ciencias del Deport. **8**, 334–356 (2012). https://doi.org/10.5232/ricyde2012.03004
10. Gochicoa-Rangel, L., et al.: Prueba de caminata de 6 minutos: recomendaciones y procedimientos. Neumol. Cir. Torax. **74**, 127–136 (2015). https://doi.org/10.1080/10520290902879763
11. Guyatt, G.H., et al.: The 6-minute walk: a new measure of exercise capacity in patients with chronic heart failure. Can. Med. Assoc. J. **132**, 919–923 (1985). https://doi.org/10.1016/j.physio.2011.11.003
12. Alonso, J., Prieto, L., Antó, J.M.: La versión española del SF-36 Health Survey (Cuestionario de Salud SF-36): un instrumento para la medida de resultados clínicos. Med Clin. **104**, 771–776 (1995). https://doi.org/10.1016/S0742-8413(99)00010-9
13. Stoet, G.: PsyToolkit: a novel web-based method for running online questionnaires and reaction-time experiments. Teach. Psychol. **44**, 24–31 (2017). https://doi.org/10.1177/0098628316677643
14. Stoet, G.: PsyToolkit: a software package for programming psychological experiments using Linux. Behav. Res. Methods. **42**, 1096–1104 (2010). https://doi.org/10.3758/BRM.42.4.1096
15. Golden, C.J.: Stroop Color and Word Test Manual (1978)
16. Golden, C.J.: A group version of the stroop color and word test. J. Pers. Assess. **39**, 386–388 (1975). https://doi.org/10.1207/s15327752jpa3904_10
17. Delfino, L.D., et al.: Screen time by different devices in adolescents association with physical inactivity domains and eating habits. J. Sports. Med. Phys. Fitness. **58**, 318–325 (2018). https://doi.org/10.23736/S0022-4707.17.06980-8
18. Mireku, M.O., et al.: Night-time screen-based media device use and adolescents' sleep and health-related quality of life. Environ. Int. **124**, 66–78 (2019). https://doi.org/10.1016/j.envint.2018.11.069
19. Biddle, S.J.H., Bengoechea García, E., Pedisic, Z., Bennie, J., Vergeer, I., Wiesner, G.: Screen time, other sedentary behaviours, and obesity risk in adults: a review of reviews. Curr. Obes. Rep. **6**(2), 134–147 (2017). https://doi.org/10.1007/s13679-017-0256-9
20. Barkley, J.E., Lepp, A.: Mobile phone use among college students is a sedentary leisure behavior which may interfere with exercise. Comput. Human Behav. **56**, 29–33 (2016). https://doi.org/10.1016/j.chb.2015.11.001
21. Fennell, C., Barkley, J.E., Lepp, A.: The relationship between cell phone use, physical activity, and sedentary behavior in adults aged 18–80. Comput. Human Behav. **90**, 53–59 (2019). https://doi.org/10.1016/j.chb.2018.08.044
22. Fang, K., Mu, M., Liu, K., He, Y.: Screen time and childhood overweight/obesity: a systematic review and meta-analysis. Child. Care. Health Dev. **45**, 744–753 (2019). https://doi.org/10.1111/cch.12701
23. Cho, Y.M., et al.: A cross-sectional study of the association between mobile phone use and symptoms of ill health. Environ. Health Toxicol. **32**, 1–7 (2016). https://doi.org/10.5620/eht.e2017001

24. Abramson, M.J., et al.: Mobile telephone use is associated with changes in cognitive function in young adolescents. Bioelectromagnetics **30**, 678–686 (2009). https://doi.org/10.1002/bem.20534
25. Wilmer, H.H., Sherman, L.E., Chein, J.M.: Smartphones and cognition: a review of research exploring the links between mobile technology habits and cognitive functioning. Front. Psychol. **8**, 1–16 (2017). https://doi.org/10.3389/fpsyg.2017.00605

The Tensiomiography Amplitude Stimulation Influences the Interpretation of the Rectus Femoris Neuromuscular Status After a Repeated Sprint Training

Alejandro Muñoz-López[1](✉), Borja Sañudo[2], and Moisés de Hoyo[2]

[1] Departamento de Motricidad Humana y Rendimiento Deportivo, Universidad de Sevilla, 41013 Sevilla, Spain
amunoz26@us.es

[2] Departamento de Educación Física y Deporte, Universidad de Sevilla, 41013 Sevilla, Spain

Abstract. This study aimed to analyze the relationship between the muscle contractile properties (MCP), before and after performing several bouts of repeated sprints, measured at different neuromuscular electrical stimulation (NMES) amplitudes. Twenty-two male participants participated on this study (age = 23.2 ± 9.8 years, height = 1.75 ± 0.01 m, body mass) 72.7 ± 21.0 kg). Testing consisted of creatine kinase (CK) measurement, tensiomyography (TMG), and maximal voluntary isometric contraction (MVIC). We first collected a fingertip capillary blood sample for the CK determination after 10 min of remaining in a seated position. Secondly, we conducted a TMG test on each limb in the Rectus Femoris (RF). Then, participants performed 6 maximal 30 m repeated sprints (RS) with a 30 s recovery. Pre-intervention, the highest correlations ($p < 0.05$) were found between muscle displacement stimulated at 40 mA and RS-change ($r_b = -0.65$), RS-Loss ($r_b = -0.66$) and RS-Fatigue ($r_b = -0.55$). Superior NMES amplitude decreased the magnitude of the correlation. After the intervention, the highest correlations were found with an amplitude of 60 mA between muscle displacement and RS-change ($r_b = -0.61$), RS-Loss ($r_b = -0.57$), and RS-Fatigue ($r_b = -0.50$). These results suggest that measuring the TMG until the Dm-max point can lead to misinterpretations of the MCP behavior in response to transient fatigue.

Keywords: Tensiomyography · Fatigue · Repeated sprint · Muscle properties

1 Introduction

The peripheral muscle fatigue implies a reduction in force-generating capacity by skeletal muscles, affecting the electrochemical and mechanical mechanisms downstream of the neuromuscular junction to the force transmission at the tendon insertion point [1]. A physical effort performed at maximum intensity can elicit a transitory muscle peripheral fatigue [2].

Tensiomyography (TMG) is a non-invasive test that measures the muscle contractile properties (MCP) in surface muscles selectively and at rest [3]. These characteristics

B. Sañudo Corrales and J. García-Fernández (Eds.): Tapasconference 2020, LNBE, pp. 43–49, 2022.
https://doi.org/10.1007/978-3-030-92897-1_5

make this tool of particular interest in the sport context, especially to measure changes after fatiguing actions, leading to a transient fatiguing state [4]. The first Tensiomyography study was published in 1996 [5]. The first protocol, which follows the manufacturer's recommendations, consists of placing a pair of surface electrodes at a minimum distance of 5 mm between, on the point of visually maximum muscle belly hypertrophy [6]. To elicit a muscle contraction, surface neuromuscular electrical stimulation (NMES) is used. The stimulation amplitude is typically increased from low NMES amplitudes (i.e., 20 mA) until the stimulation amplitude in which the muscle elicits its' maximum displacement (Dm). The last is a typical protocol that has been extended to most of the research to date, to our knowledge.

The prior protocol lies on the base that the higher the stimulation amplitude, the higher the number of motor units (MU) can be activated [7]. Other passive neuromuscular properties are studies at the maximum Dm (Dm-max), such as the muscle contraction time (Tc). However, Loturco et al. [8] found a better prediction of elite sprint performance when a submaximal amplitude of 40 mA was used. Henneman's size principle reads that smaller MU (mainly composed of Type I muscle fibers) is first activated when a muscle is voluntarily activated. In contrast, bigger MU is secondly recruited. Nevertheless, during NMES, the MU recruitment threshold depends on the geometric factor, the uniform distribution of MU types and the depth of the axonal branches, which innervates each kind of MU type [9]. In combination, Jubeau et al. [10] found that using NMES compared to voluntary contractions, MU are activated without evident sequencing related to different types.

Hence, we hypothesized that the fast MU are firstly recruited when an incremental stimulation amplitude is used. In addition, Type II fibers are easily fatigued compared to Type I fibers[11]. While most of the research analyzes the MCP at the point of Dm-max (theoretically the maximum MU possible recruitment using NMES), some authors found interesting relationships using lower NMES intensities [8]. Therefore, the MCP acute responses to fatiguing training can be related to different stimulation intensities. However, to our knowledge, this has not been studied yet. This study aimed to analyze the relationship between the MCP before and after several repeated sprints, measured at different NMES amplitudes, and the task performance and fatigue markers. We hypothesized that performing repeated sprints will show transient changes in MCP. Those changes will depend on the NMES amplitude used during the assessment in relation to other performance and fatigue markers.

2 Methods

2.1 Study Design

This study follows a cross-sectional study. Participants performed a familiarization a week before the training intervention. Testing consisted of CK measurement, TMG, and maximal voluntary isometric contraction [MVIC]. We first collected a fingertip capillary blood sample for the CK determination after 10 min of remaining in a seated position. Secondly, we conducted a TMG test on each limb in the Rectus Femoris (RF). Then, participants performed a warm-up consisting of 3 min of joint mobilization exercises and 5 min of cycling at a comfortable pace (80w at 100 rpm) on an electronically-braked cycle ergometer (Ergoselect 200, Ergoline, Germany). Finally, participants executed the MVIC test. The training intervention consisted of 6 maximal 30 m repeated sprints (RS) with a 30 s recovery period between each effort (Mujika et al., 2009). We finally conducted the same pre-testing battery test immediately after the last sprint.

2.2 Participants

Twenty-two male participants participated on this study (age = 23.2 ± 9.8 years, height = 1.75 ± 0.01 m, body mass) 72.7 ± 21.0 kg). They were physically active athletes who frequently took part in intermittent sports and represented their respective sports at either university, county, or national levels. Participants were asked not to alter their regular lifestyle habits and diet throughout the study. They were required to refrain from exercise, caffeine, and alcohol for 24 h before testing and to abstain from vigorous exercise throughout the experimental period. Each participant was fully informed about procedures, potential risks, and benefits of the study, and they all signed written informed consent before the tests. The study was conducted following the Declaration of Helsinki and was approved by a local ethics committee.

2.3 Repeated Sprints

Before the RS protocol, participants underwent a specific sprinting warm-up consisted of four 20-m running progressive sprints in intensity with one minute between them. Participants stayed standing start with the lead-off foot placed 1 m away from the first timing gate (with a vertical height of 0.80 m). Then, participants completed 6 × 30 m maximal sprints with 30-s rest between sprints on an indoor running track marked using four sets of timing gates (Polifemo Radio Light; Microgate, Bolzano, Italy). Standardized verbal encouragements were given throughout the protocol. To account for the RS performance, we calculated the best RS time trail (RS-Best), the loss in RS time (RS-Loss) as the difference between the first and last time trial, and the RS-Fatigue [12].

2.4 Creatine Kinase Concentration

Plasma CK concentration was assessed at all time points tests from 30-μL capillarized whole blood samples collected via fingertip puncture made using a spring-loaded disposable lancet (Safe-T-Pro Plus, Accu-Chek, Roche Diagnostics GmBH, Germany). The whole blood sample was immediately pipetted to a test strip and analyzed for CK concentration using a colorimetric assay procedure (Reflotron Plus; Roche Diagnostics, West Sussex, United Kingdom).

2.5 Maximal Voluntary Contraction

We tested the MVIC of the knee extensors in both legs using an isokinetic dynamometer (Biodex System 4, Biodex Medical Systems, Shirley, NY), sampled at 100 Hz. Each participant remained seated at an angle of 85 ° and stabilized with straps at the shoulders, waist, and thighs as per the manufacturer's guidelines. Participants performed three maximal attempts for each of the isokinetic contractions at an angle of 90°, separated by 60 s. All participants were given verbal encouragement and visual feedback of the torque signal at each repetition. We chose the peak torque value (N·m), obtained from the highest value of all maximum efforts, for each muscle action was chosen for further analysis.

2.6 Tensiomyography

MCP were assessed using TMG in the RF of both legs. Tc and Dm were analyzed, as those variables are reproducible [13] and susceptible to show transient acute fatigue after a RS protocol [12]. The general protocol followed the manufacturer's recommendations described elsewhere [4]. Specifically, NMES amplitude started at 20-mA and was increased in ranges of 20-mA until 100-mA, every 15 s to avoid electrical fatigue of the motor unit [13]. Hence, a total of 5 stimulations were used. All the values from Tc and Dm were stored for subsequent analyses. The sensor location was determined anatomically at the point of maximal muscle response [14] and marked with a dermatological pen for all the posterior measurements. Participants were asked to maintain marks during the intervention process.

2.7 Statistical Analyses

Data are shown as mean ± standard deviation. Normality assumption was tested before any statistical test using the Shapiro-Wilk test. We performed a correlation analysis to study the possible relationships at Pre and Post between MCP and RS performance or Fatigue markers, using the Spearman raking (r_b, non-normal distribution). Secondly, we analyzed the differences between pre-intervention to post-intervention, using a test for paired samples (T-test for normal distributions or Wilcoxon signed-rank for non-normal distributions). In addition, we calculated effect sizes (ES) for the repeated measures designs using the Partial eta-squared ($\eta^2{}_p$), in addition to Cohen's d-effect (d) or rank-biserial correlation (r_b) size statistics for parametric and non-parametric paired comparisons, respectively. For all the analyses, we set the significant level at $p < 0.05$. We used the Rstudio software (v. 1.2.5033, RStudio: Integrated Development for R. Rstudio, Inc., Boston, MA) with the corrplot package (v. 0.84) to perform the correlation analyses and correlation graphs, and the JASP software (v. 0.14.1, JASP Team) for the rest of the analyses.

3 Results

The Dm-Max was found at an average amplitude of 86.35 ± 23.97 and 91.49 ± 19.05 mA in the RF, at basal and post, respectively. There were significantly different correlations in relation to the NMES amplitude used (Fig. 1). Pre-intervention, the highest correlations ($p < 0.05$) were found between muscle displacement stimulated at 40 mA and RS-change (r_b = –0.65), RS-Loss (r_b = –0.66) and RS-Fatigue (r_b = –0.55). Superior NMES amplitude decreased the magnitude of the correlation, despite being significant. After the intervention, the highest correlations were found with an amplitude of 60 mA

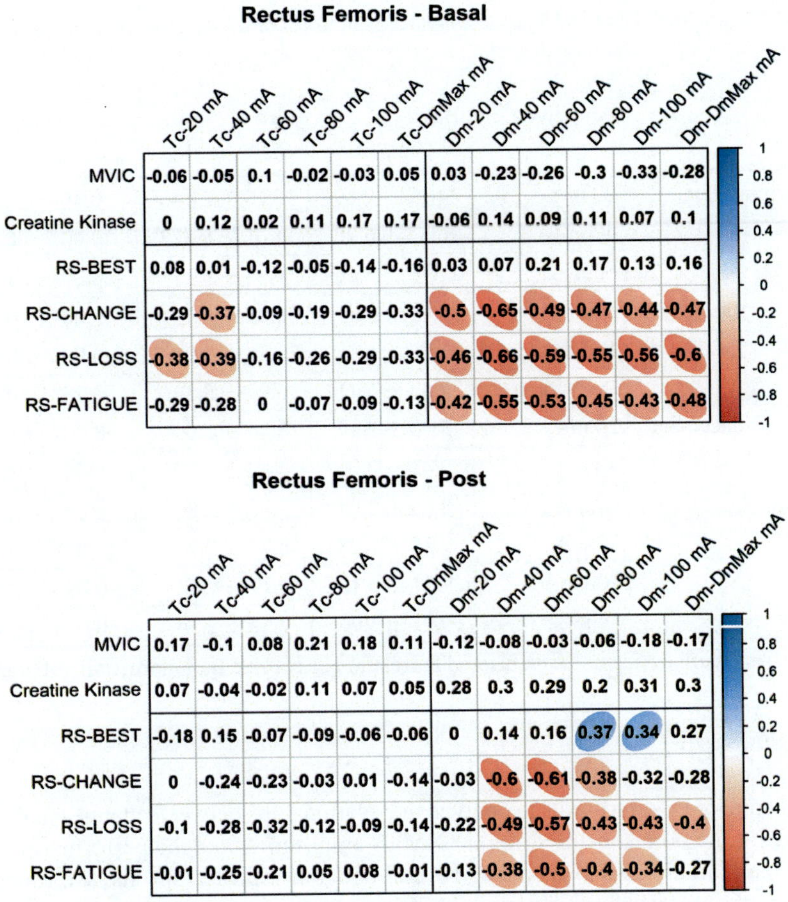

Fig. 1. The Spearman correlation coefficients between contraction time (Tc), or muscle displacement (Dm), and RS-Performance and Fatigue markers. Significant correlations ($p < 0.05$) are flagged as a colored circle (reds for negative and blues for positive correlations).

between muscle displacement and RS-change ($r_b = -0.61$), RS-Loss ($r_b = -0.57$), and RS-Fatigue ($r_b = -0.50$).

Table 1 shows the differences between Pre- and Post-intervention at all the NMES amplitudes for Tc and Dm. Tc showed significant differences when amplitudes of 40, 60, 80, and 100 mA were used. In contrast, Dm showed significant differences at all the amplitudes. In summary, while Tc showed a significant increase, Dm showed a significant decrease, except when 20 mA were used (significant increase). The biggest changes in Tc and Dm were observed with the lower NEMS amplitudes.

Table 1. Contraction Time and Muscle Displacement time course descriptive values a time effect changes

Muscle	Amplitude	Basal (mean ± SD)	Post (mean ± SD)	Time effect	
				p-value	ES
Contraction Time	20 mA	34.12 ± 9.69	36.18 ± 11.46	0.469	0.126
	40 mA	26.57 ± 6.28	34.61 ± 8.57	<0.001	0.721
	60 mA	27.73 ± 5.86	31.94 ± 5.77	<0.001	1.056
	80 mA	27.07 ± 4.37	28.22 ± 6.49	<0.001	0.661*
	100 mA	26.37 ± 4.14	27.59 ± 3.96	0.003	0.612*
	DmMax mA	27.34 ± 5.11	26.99 ± 3.83	0.702	0.081*
Muscle displacement	20 mA	3.33 ± 2.16	7.90 ± 3.63	<0.001	1.326
	40 mA	7.42 ± 3.00	4.32 ± 2.77	<0.001	1.112
	60 mA	8.65 ± 2.88	7.90 ± 2.86	0.005	0.550*
	80 mA	9.25 ± 2.98	8.73 ± 2.70	0.031	0.454*
	100 mA	9.77 ± 3.12	9.17 ± 2.62	0.021	0.454*
	DmMax mA	10.01 ± 3.09	9.52 ± 2.76	0.043	0.400*

ES = Cohens'd effect size or *rank biserial correlation effect size (non-normal distribution).

4 Discussion

The objective of this study was to analyze the relationships between changes in the MCP at different time points, using different NMES amplitudes, and the performance during and fatigue markers after performing several bouts of repeated sprints. We found that, in the RF, the relationships between MCP and performance or fatigue markers were different in relation to the NMES amplitude used. More concretely, while no significant relationships were found between MCP and fatigue markers (i.e., MVIC or CK), the highest correlations with RS Performance were found at 40 mA with RS-Change, RS-Loss, and RS-Fatigue at a basal time point, but at 60 mA with RS-Best, RS-Change, RS-Loss and RS-Fatigue at post-exercise time point. What is more, Tc was different between Basal and Post at 40, 60, 980, and 100 mA, while Dm was different at any NMES amplitude used.

The assessment of muscle fatigue has been a topic of interest for years in sports science research [15]. The TMG is sensible enough to detect changes in the MCP at rest, showing acute transient fatigue after exercise [4, 16]. In agreement with Loturco et al. [17], we found that a submaximal NMES amplitude (i.e., 40 mA) showed the highest correlations at basal level with RS Performance. In addition, we also found that at post, another submaximal NMES amplitude (i.e., 60 mA) showed the highest relationships. In contrast, the maximum Dm was found at NMES intensities between 80 and 100 mA. These results suggest that measuring the TMG until the Dm-max point can lead to misinterpretations of the MCP behavior in response to transient fatigue. A possible explanation of why a lower NMES showed higher relationships with performance, even at post level, is that the fast MU are firstly activated with NMES and, of note, are the most fatiguing fibers. Hence, a lower NMES amplitude is required to activate all the MU, which can better explain both performance and fatigue in sport, the fast MU [11]. This is of practical importance because most of the research used the Dm-max point to assess the MCP at rest.

References

1. Cè, E., Longo, S., Limonta, E., Coratella, G., Rampichini, S., Esposito, F.: Peripheral fatigue : new mechanistic insights from recent technologies. Eur. J. Appl. Physiol. **120**(1), 1739 (2019)
2. Sánchez-Sánchez, J., et al.: Effect of a repeated sprint ability test on the muscle contractile properties in elite futsal players. Sci. Rep. **8**(1), 1–8 (2018)
3. Simunic, B.: Tensiomyographic Contraction Time of Selected Skeletal Muscles (2011)
4. Muñoz-López, A., De Hoyo, M., Nuñez, F.J., Sañudo, B.: Using tensiomyography to assess changes in knee muscle contraction properties after concentric and eccentric fatiguing muscle actions. J. Strength Cond. Res. 37–41 (2020). (Epub ahead(February))
5. Burger, H., Valenčič, V., Marinček, Č, Kogovšek, N.: Properties of musculus gluteus maximus in above-knee amputees. Clin. Biomech. **11**(1), 35–38 (1996)
6. Dahmane, R., Valenčič, V., Knez, N., Eržen, I.: Evaluation of the ability to make non-invasive estimation of muscle contractile properties on the basis of the muscle belly response. Med. Biol. Eng. Comput. **39**(1), 51–55 (2001)
7. Stratton, K., Faghri, P.D.: Electrically and hybrid-induced muscle activations: effects of muscle size and fiber type. Eur. J. Transl. Myol. **26**(3), 249–254 (2016)
8. Loturco, I., et al.: Predictive factors of elite sprint performance: influences of muscle mechanical properties and functional parameters. J. strength Cond. Res. **33**(4), 974–986 (2019)
9. Knaflitz, M., Merletti, R., De Luca, C.J.: Inference of motor unit recruitment order in voluntary and electrically elicited contractions. J. Appl. Physiol. **68**(4), 1657–1667 (1990)
10. Jubeau, M., Gondin, J., Martin, A., Sartorio, A., Maffiuletti, N.A.: Random motor unit activation by electrostimulation. Int. J. Sports Med. **28**(11), 901–904 (2007)
11. Bagley, J.R., McLeland, K.A., Arevalo, J.A., Brown, L.E., Coburn, J.W., Galpin, A.J.: Skeletal muscle fatigability and myosin heavy chain fiber type in resistance trained men. J. Strength Cond. Res. **31**(3), 602–607 (2017)
12. Sánchez-Sánchez, J., et al.: Repeated sprint ability and muscular responses according to the age category in elite youth soccer players. Front. Physiol. **10**(12), 1–8 (2019)
13. Križaj, D., Šimunič, B., Žagar, T.: Short-term repeatability of parameters extracted from radial displacement of muscle belly. J. Electromyogr. Kinesiol. **18**(4), 645–651 (2008)
14. Djordjevi, S., Tomai, S., Zupani, G., Piot, R., Dahmane, R.: The influence of different elbow angles on the twitch response of the biceps brachii muscle between intermittent electrical stimulations. In: Applied Biological Engineering- Principles and Practice no. 1994 (2012)
15. Mendez-Villanueva, A., Hamer, P., Bishop, D.: Physical fitness and performance. Fatigue responses during repeated sprints matched for initial mechanical output. Med. Sci. Sports Exerc. **39**(12), 2219–2225 (2007)
16. Giovanelli, N., Taboga, P., Rejc, E., Simunic, B., Antonutto, G., Lazzer, S.: Effects of an uphill marathon on running mechanics and lower-limb muscle fatigue. Int. J. Sports Physiol. Perform. **11**(4), 522–529 (2016)
17. Loturco, I., et al.: Muscle contraction velocity: a suitable approach to analyze the functional adaptations in elite soccer players. J. Sport. Sci. Med. **15**(3), 483–491 (2016)

Innovation in Sports and Health

Benefits of Virtual Reality in People with Chronic Pain

Ana Belén Parra Díaz[1(✉)], F. Hita-Contreras[1], A. Martínez-Amat[1], María del Carmen Carcelén Fraile[1], R. Fabrega-Cuadros[1], J. D. Jiménez-García[2], María Alzar Tcrucl[1], and A. Aibar-Almazan[2]

[1] Department of Health Sciences, Faculty of Helath Sciences, University of Jaén, 23071 Jaén, Spain
[2] MOVE-IT Research Group, Department of Physical Education, Faculty of Education Sciences, University of Cádiz, Cádiz, Spain
josedaniel.jimenez@uca.es

Abstract. Virtual reality is a technological strategy that introduces the person into a fictional and interactive world, causing the subjects to evade reality and it is used in the clinical setting in order to reduce pain. The purpose of this literature review is to investigate the benefits of virtual reality in people with chronic pain. A search was carried out in the Pubmed, CINAHL and Web of Science databases, using the following terms: "Virtual reality" AND "Chronic pain". Those articles that reported on randomized controlled trials published since 2010 were chosen, discarding duplicate articles and those that did not meet the eligibility criteria. Of the 231 articles identified in the search, 9 articles were finally included, with a total of 368 people who participated in the studies. All articles were aimed at testing the effects of virtual reality on chronic conditions. We conclude that virtual reality helps to control pain, although more studies are needed to provide evidence, especially regarding the long-term effects of this therapy in chronic pain.

Keywords: Fall risk · Falls efficacy scale international · Gait speed · Elderly · Older people

1 Introduction

Chronic pain is defined as pain that lasts more than 3 or 6 months from its onset or lasts longer than necessary for its healing [1]. Approximately 17% of the Spanish population suffers from chronic pain [2], which is the cause that causes the most disability worldwide and is suffered by 30% of the world's population [3, 4]. It mainly affects women and is associated with degenerative diseases and age[4]; around 50–80% of people over 65 years of age have chronic pain [5].

Chronic pain is a public health problem due to its high prevalence [6], since it requires many health resources such as medical consultations and the use of drugs, and it also generates a high psychological and social impact since it becomes disabling.

The treatment for pain is mainly pharmacological; the most used drugs are opioids, so it is very important to address pain with other treatments and therapies that do not generate

B. Sañudo Corrales and J. García-Fernández (Eds.): Tapasconference 2020, LNBE, pp. 53–58, 2022.
https://doi.org/10.1007/978-3-030-92897-1_6

dependency or side effects [7]. Ideally, these people should have a multidisciplinary treatment in which pharmacological treatment is combined with non-pharmacological treatments [8]. For the approach to chronic pain, the most used non-pharmacological therapies are physical exercise, rehabilitation treatment and psychological therapies.

New technologies are very important for the diagnosis and treatment of multiple clinical conditions. Virtual Reality (VR) is a non-pharmacological therapy that in recent years is being used for the treatment of chronic pain, it is a technological therapy that consists of transporting subjects to a virtual world in which they interact in an environment in which can escape pain [8, 9].

Virtual reality is mainly used as a distraction technique to relieve pain, since through this activity the intensity and discomfort caused by pain are modified by changing the central axis of attention [10], so VR can be used as a treatment for ease the pain. The objective of this study is to investigate the benefits of virtual reality in people with chronic pain.

2 Methods

A literature review was carried out in the Pubmed, CINAHL and Web of Science databases, using the following search terms: "Virtual reality" AND "Chronic pain". Those articles that were randomized controlled trials published from 2010 to 2020 were chosen, and duplicate articles and those that did not meet the eligibility criteria were discarded.

3 Results

Of the 231 articles identified in the search, 9 articles were finally included, with a total of 389 people participating in the studies. All the articles aimed to verify the effects of virtual reality in chronic pain.

In Table 1, you can see data obtained in the analyzed articles in which the study, number of participants, objectives and results obtained in each of the studies are represented. As can be seen in all the studies, virtual reality reduces pain during use [8, 9, 11–17] and also improves emotional aspects. However, physical exercise is more beneficial than virtual reality in reducing pain and functionality in people suffering from chronic pain according to the study carried out by Morales et al. [17].

Table 1. Data obtained in the articles analyzed on the use of virtual reality in chronic pain.

Study	Participants	Objectives	Results
Cebas, 2011	n = 72	To know the results of distracting VR on pain-related cognitions	Significant increase in pain tolerance, increase in the self-efficacy component, and decrease in despair and magnification
Loreto, 2011	n = 90	Knowing the correspondence between chronic pain induced in the laboratory and an enteroscopic figure, which represents the sensation of pain using virtual reality	Increase in pain tolerance, pain threshold and self-efficacy in pain management and decrease in catastrophic thoughts
Herrero et al. 2013	n = 8	To present the development of a VR system to increase positive emotions in women with fibromyalgia	Significant reduction in pain, increased sense of efficiency and increased sense of energy and vigor
Garrett et al., 2020	n = 23	Know the experiences of virtual reality in the pain of people with cancer	Efficacy of VR in relieving pain during use
Aguirre, 2015	n = 77	To evaluate the effect of ipsilateral upper limb exercises performed with VR from the seventh day in women undergoing mastectomy, evaluate the presence of pre and post-operative pain, determine the degree of functionality and perform mobilization activities through exercises supported by VR from the seventh day of the operation	Greater reduction in pain and increased functionality of the patients' upper limb
Jones et al., 2016	n = 30	To determine the impact of a VR session on the experience of pain in patients with chronic pain	Participants reduced their pain in half during the VR session
Flujas et al. 2017	n = 6	Show a psychological treatment program for pain and anxiety and emotional discomfort in children and adolescents in hospitalized children and adolescents through the use of ICTs	High score in all variables, this treatment improves bad emotional anxiety and pain in children and adolescents

(*continued*)

Table 1. (*continued*)

Study	Participants	Objectives	Results
Steward et al. 2019	n = 128	Knowing the impact of VR on pain in people hospitalized with heart failure. Evaluate the acceptance and satisfaction of the hospitalized patient with heart failure	Virtual reality is effective together with pharmacological treatment to control pain in people hospitalized with heart failure
Morales et al. 2020	n = 44	To compare the effects of virtual reality with the effects of physical exercise and functional, somatosensory and functional variables in the treatment of chronic neck pain	Physical exercise was found to be better than virtual reality in improving pain reduction, neck functionality, and feelings of catastrophe and fear

Note: n = number of participants

4 Discussion

In the studies analyzed, virtual reality improves pain control, in agreement with these results are Miró et al. [10], who after conducting a systematic search on virtual reality and chronic pain came to the conclusion that VR is effective in reducing pain, sensory and emotional factors of pain. Mallari et al. [18]. also agree with the benefits of virtual reality in pain after having carried out a review of studies between virtual reality and acute and chronic pain in which they verified that acute pain was reduced during activity and after and chronic pain also decreased in the virtual reality sessions and immediately after, but they could not demonstrate the long-term results of this type of therapy.

Psychological aspects play a very important role in pain management, for this reason virtual reality is used as a behavioral psychology therapy that uses attentional distractors to avoid paying attention to pain and reducing it. Based on this type of behavioral intervention, Flujas et al. [16], carried out a psychological intervention program and virtual reality in hospitalized children and adolescents with chronic diseases to improve pain, anxiety and emotional distress, showing certain improvements in pain regulation and anxiety management. These data are also supported by Zhan et al. [2], who investigated the use of technology for chronic pain as a distraction and feedback medium in amputated phantom limb and observed that during the time that VR treatment was being carried out to train motor and sensory functions decreased pain and improved up to a week after finishing the intervention, but after this period the phantom limb returned to pain as before starting the sessions.

Regarding therapies based on new technologies, according to Jones et al. [12], virtual reality obtains better results for pain relief than other technological therapies such as video games, because it creates very real environments that can be transplanted into real life, the same is the opinion Garrett et al. [13], Who carried out a study of new technologies in cancer patients with chronic pain in which they used virtual reality and online therapies and the patients agreed that with VR their pain improved although they found limitations of accessibility and management at the time to access virtual reality.

All studies indicate that virtual reality improves pain control and is an effective therapy to combine with pharmacological treatment. In addition to improving and reducing pain, it also has benefits on a psychological level, since it helps to control situations of stress and emotional discomfort and makes subjects feel more vital and eager to face tasks, improving functionality to carry out daily activities that are sometimes disabling due to pain. However, more studies are needed to provide evidence to the scientific literature, since there is little information on the long-term effects of this therapy in chronic pain.

According to these results, virtual reality has a positive effect on pain treatment but has limitations when accessing it, for economic, learning and management reasons, especially in the elderly population.

References

1. Covarrubias, A.: Papel del anestesiólogo en el manejo del dolor crónico. Rev. Mex. Anestesiol **31**(1), 235–237 (2008)
2. Torralba, A., Miquel, A., Darba, J.: Situación actual del dolor crónico en España: Iniciativa "Pain Proposal". Rev. Soc. Esp. Dolor **21**(1), 16–22 (2014)
3. Zhang, S., Seymour, B.: Technology for chronic pain. Curr. Biol. **24**(18), 930–935 (2014)
4. Bautista, S.G., Jiménez, A.: Epidemiologia del dolor crónico. Arch. Med. Familiar **16**(4), 69–72 (2014)
5. Franco, M.L., de Lucas, S.: A Características del dolor crónico en el anciano: tratamiento. Rev. Soc. Esp. Dolor. **8**, 28–29 (2001)
6. Cantero, L.M., Cabrera, A., López, M.A., Saúl, L.A.: Intervención grupal desde un enfoque sensoriomotriz para reducir la intensidad del dolor crónico. Aten. Primaria **51**(3), 162–171 (2019)
7. Honzel, E., et al.: Virtual reality, music and pain: developing the premise for an interdisciplinary approach to pain managent. Pain. Author Manuscript **160**(9), 1909–1919 (2020)
8. Steward, D., Mete, M., Groninger, H.: Virtual reality for pain management in patients with heart failure: study rationale and design. Contemp. Clin. Trials. Commun. **16**, 100470 (2019)
9. Herrero, R., Castilla, D., Vizcaino, Y., Molinari, G., Garcia-Palacios, A., Botella, C.: Avances en el tratamiento psicológico de la fibromyalgia: el uso de la realidad vistual para la inducción de emociones positivas y la promoción de la activación comportamental. Un estudio piloto. Rev. Arg. clín. psicol. **2**, 111–120 (2013)
10. Miró, J., Nieto, R., Huguet, A.: Realidad virtual y manejo del dolor. C. Med. Psicosom. **82**, 52–64 (2007)
11. Cebas, K.: Utilización de la RV distactiva para cambiar las cogniciones relacionadas con el dolor. Revistaesalud.com. **7**(26), 3–4 (2011)
12. Loreto, D.: uso de la imagen guiada en RV para cambiar las cogniciones relacionadas con el dolor. Revistaesalud.com. **7**(26), 5–6 (2011)
13. Garrett, B.M., Tao, G., Taverner, T., Cordingley, E., Sun, C.: Patients perception of virtual reality therapy in de management of chronic cáncer pain. Heliyon. **6**, 1–10 (2020)

14. Aguirre, M., Marchant, P.: Descripción del efecto de los ejercicios de la extremidad superior ipsilateral realizados con realidad virtual en mujeres sometidas a mastectomía. Gaz. Mex. Onc. **14**(4), 204–209 (2015)
15. Jones, T., Moore, T., Choo, J.: The impact of virtual reality on chronic pain. PLoS ONE **20**, 1–10 (2016)
16. Flujas, J.M., Ruíz, D., Botella, C., Gómez, I.: Un programa de bienestar emocional basado en Realidad Virtual y Terapia Online para enfermedades crónicas en infancia y adolescencia: La academia espacial. Rev. Psicol. Clin. niños y adolescentes **4**(3), 17–25 (2017)
17. Morales, D., et al.: Effects of virtual reality versus exercise on pain, functional, somatosensory, psychosocial outcomes in patients whit non-specific chronic neck pain: a radomized clinical trial. Int. J. Environ. Res. Public Health **17**, 1–19 (2020)
18. Mallari, B., Spaeth, E.K., Gog, H., Boyd, B.S.: Virtual reality as analgesic for acute and chronic pain in adults: a systematic review and a meta-analysis. J. Pain Res. **12**, 2053–2085 (2019)

Exergaming as a Neurorehabilitation Tool in Patients Diagnosed with a Severe Mental Disorder: A Review of Current Scientific Evidence

Daniel Gallardo Gómez(✉)

Universidad de Sevilla, Sevilla, Spain
dggomez@us.es

Abstract. Severe mental disorders are generally characterized by a combination of abnormal thoughts, perceptions, emotions, behavior, and relationships with others, often leading to decreased quality of life and increased anxious and depressive symptoms. A tool that is beginning to be used with this group is exergaming, with the objective to improve their general health levels. The main aim of this research is to report exergaming intervention effects in people diagnosed with a severe mental illness. An in-depth database search and selection criteria were applied for study eligibility. The results of the four included studies showed that exergames-based interventions had positive effects on quality of life and depressive symptoms comparing with control conditions. These effects did not differ a lot when these interventions were compared with active controls or more traditional interventions. In conclusion, the evidence supports the use of exergames to improve quality of life and depressive symptoms in patients diagnosed with a severe mental disorder. Nevertheless, it is a poorly studied population, and more research is needed to provide adequate specific guidelines and recommendations on virtual reality-based physical activity programming.

Keywords: Severe mental disorder · Neurorehabilitation · Exergaming · Quality of life

1 Introduction

According to the World Health Organization (WHO), severe mental disorders are generally characterized by a combination of abnormal thoughts, perceptions, emotions, behavior, and relationships with others. These characteristics make severe mental disorders one of the world's leading causes of disability, which leads to a marked decrease in quality of life. These patients are also prone to developing anxious and depressive symptoms (Wang et al. 2007). The latest statistical data on the incidence of SMD in the world population show that one in ten persons suffers one of them (10.7%). All of this makes the management and treatment of severe mental disorders a necessity (James et al. 2018).

B. Sañudo Corrales and J. García-Fernández (Eds.): Tapasconference 2020, LNBE, pp. 59–65, 2022.
https://doi.org/10.1007/978-3-030-92897-1_7

Nowadays, the main treatment for these patients is based on antipsychotic medication in most cases, although the use of alternative therapies such as psychotherapy, cognitive or group therapies are recommended to especially reduce depressive and affective symptoms (Huhn et al. 2014). An alternative therapy that is growing in recent years is programmed physical activity, showing beneficial effects on psychopathology, physical and cognitive functions of these patients (Vancampfort et al. 2017). Due to the development of new technologies and the adaptation of physical activity programs due to the current pandemic situation, new forms of physical activity have been developed, such as exergames (to do physical exercise with virtual reality systems).

Current evidence supports the use of physical activity for the prevention and management of many serious mental illnesses (e.g., depression, schizophrenia, anxiety, etc.) (Deenik et al. 2017; Vancampfort et al. 2017), but very few studies have attempted to elucidate the effects of exergames in this population. The little existing evidence supposes that this practice of physical exercise based on virtual reality may have beneficial effects on the physical and mental health of these patients (Dellazizzo et al. 2020; Swinnen et al. 2020).

Thus, based on the scarce and low-quality evidence, this systematic review seeks to qualitatively show the effects that the exergame-based interventions have on quality of life and depressive symptoms in patients suffering from a severe mental disorder.

2 Methods

2.1 Systematic Search

A systematic search was performed using seven electronic databases: PubMed, SportDiscus, Cumulative Index to Nursing & Allied Health Literature (CINAHL), Scopus, Web of Science (WOS), PsycINFO and the Cochrane Library, in September 2020.

We selected controlled and uncontrolled terms to develop the search strategy, which was designed to respond to our acronym PICO. The terms used in searches are described below:

- Terms related to patients diagnosed with severe mental disorder: 'severe mental disorder patients', 'SMD patients', 'developmental disorder', 'schizophrenia disorder patient', 'dementia patient', 'bipolar disorder patient'.
- Terms related to physical activity and physical exercise through virtual reality (VR) and videogames: 'exergames', 'exergaming', 'physical video-games', 'virtual reality games'.
- Terms related to outcomes: 'quality of life', 'life quality', 'well-being', 'anxiety', 'depression symptoms'.
- Terms related to the type of the studies: 'systematic review', 'meta-analysis', 'meta-review'.

2.2 Eligibility Criteria for Selecting Studies

To reduce the risk of bias, we establish the following eligibility criteria: (1) systematic reviews with or without meta-analysis or reviews of meta-analysis (2) in English (3) that have studied the effects of interventions based on physical exercise through the application of virtual reality or video-games (4) on quality of life or depressive symptoms (5) in patients diagnosed with a severe mental disorder (6) by the tenth or eleventh edition of the International Classification of Diseases (ICD-10 or ICD-11) or the fourth or fifth edition of the Diagnostic and Statistical Manual of Mental Disorders (DSM-IV or DSM-V).

3 Results

3.1 Included Studies

After applying the eligibility criteria, we obtained 271 possible studies. After reviewing titles and abstracts, we obtained 28 possible studies. After reviewing full-texts, 4 studies were finally obtained that met all the criteria. The entire screening process can be seen in the Fig. 1.

One study is a meta-review of meta-analyses (Dellazizzo et al. 2020), other study is a meta-analysis (Fodor et al. 2018) and two studies are systematic reviews (Maples-Keller et al. 2017; Rus-Calafell et al. 2018). Publication year of the included studies varied between 2017 and 2020. The characteristics of the included studies are very heterogeneous, with age ranges between 18 and over 80 years, and with highly varied populations. These studies collected a wide variety of severe mental disorders such as anxiety disorders, trauma-and-stressor-related disorders, depressive disorders, schizophrenia spectrum disorders, severe-autism disorders, and panic-phobias disorders.

3.2 Outcomes

Quality of Life

Only one study analyzed the quality of life outcome directly (Maples-Keller et al. 2017). The other studies (Fodor et al. 2018; Rus-Calafell et al. 2018; Dellazizzo et al. 2020) reported improvements in physical function, social skills, cognitive function, negative illness symptoms (anxiety or psychotic symptoms) and positive symptoms in psychotic disorders such as the schizophrenia spectrum disorders (visual and auditory hallucinations), which indirectly causes improvements in the quality of life of these patients.

Maples-Keller et al. (2017) study does not provide a quantitative measure of the effects of physical activity with virtual reality (i.e., effect size) because of the type of research it is (systematic review). The improvements on quality of life occurred when virtual reality interventions were compared with pure or active control. However, when they were compared with other traditional interventions such as psychotherapy or social skills group therapy, there were no significant differences on the reported benefits.

Depressive Symptoms

The four included studies supported the efficacy of this type of interventions for the treatment of depressive symptoms in patients with severe mental disorder.

Three of the reviews included in this study (Maples-Keller et al. 2017; Rus-Calafell et al. 2018; Dellazizzo et al. 2020) endorsed physical activity based on exergames as

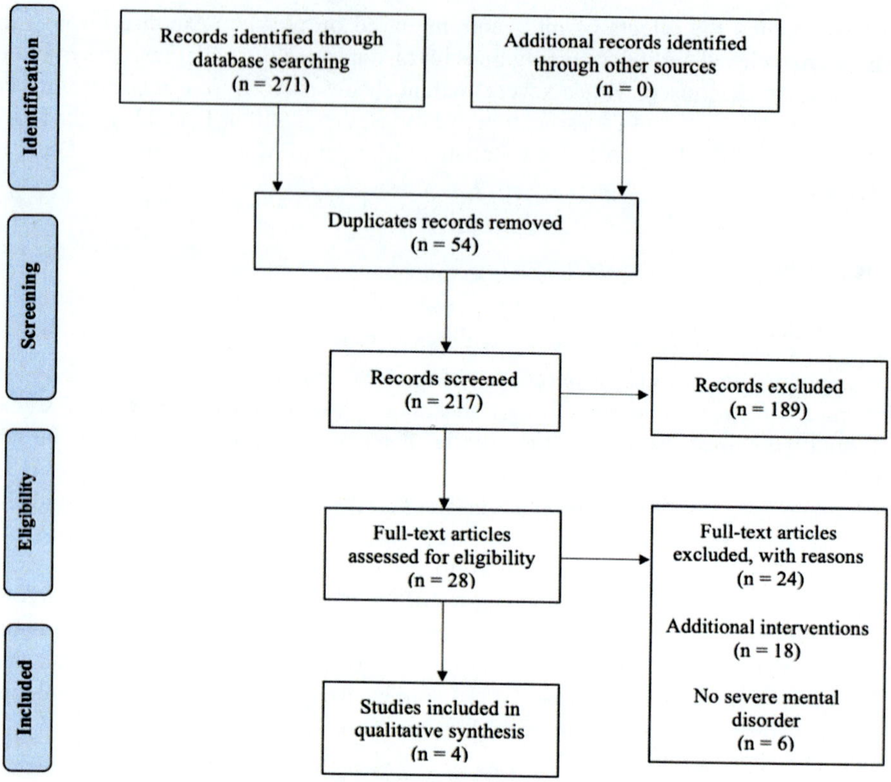

Fig. 1. PRISMA flow diagram of the screening process

one of the best alternative therapies to alleviate depressive symptomatology, which it is characteristic of most severe mental illnesses.

In Fodor et al. study (2018) a large effect size (Cohen 2013) was reported for the exergame intervention compared to pure control conditions (Hedges' $g = 0.73$, [95% Confidence Interval: 0.25, 1.21]). However, when making the comparison with groups that performed active control, there was no such effect.

4 Discussion

This review of the current scientific evidence has attempted to qualitatively report the effects that interventions based on physical activity or exercise have using new technologies (virtual reality or videogames) on quality of life and depressive symptoms in patients diagnosed with severe mental disorders.

Low to moderate quality scientific evidence showed that interventions based on exergaming or virtual reality are a tool to be considered for improving quality of life and depressive symptoms in the population suffering a severe mental disorder. These improvements in the general health of these patients may be associated with the stimulation of neurotransmitters such as the brain-derived neurotrophic factor (BDNF), which is very important in brain and cognitive functioning (Piepmeier and Etnier 2015). In addition, the optimal stimulus for BDNF production seems to be associated with moderate intensity exercise, which the exergaming mostly meets in these specific patients (most studies reported intensities between 40–70% of maximum heart rate). The combination of physical exercise itself and the stimulation of BDNF helps to produce the process of neurogenesis, essential to reduce, even partially reverse, the symptoms of some severe mental disorders, especially in those in which there is a direct deterioration of the cerebral cortex such as dementia (Weinstein et al. 2014) or Alzheimer's disease (Romito-Di Giacomo et al. 2007).

However, the effects of the studied interventions do not differ much from those caused by more traditional interventions such as psychotherapy or social skills group therapies. This may be due to the fact that physical exercise is a therapy that is not integrated in the treatment of this type of disorders, so it is less studied than others. In the last years, physical therapy has been used as part of the treatment of serious mental disorders such as schizophrenia or depressive disorder, although virtual reality and exergaming are modalities that are being incorporated into treatment very recently (Välimäki et al. 2014).

Physical therapy through exergaming or virtual reality could be an effective treatment attached to the pharmacological, which must be considered, apart from the effects on quality of life and depressive symptoms, due to its psychosocial impact on patients: stress reduction, increased energy, mood and self-esteem improvements and increased motivation (Firth et al. 2016). The motivational factor of the exergames plays a fundamental role in the management of severe mental disorders, since in most cases there is a total loss of motivation. Besides a characteristic that can support the use of this interventions in patients with depressive disorder may be that these types of patients often have associated fears and phobias. Virtual reality can simulate in a very real way the situations that help to overcome those fears (Dellazizzo et al. 2020).

A powerful reason why the application of interventions based in exergames is recommended is because they can alleviate the side effects associated with taking antipsychotic drugs in patients with psychotic disorders, belonging to severe mental disorder. These side effects usually refer to the development of cardiometabolic disorders such as diabetes, hypertension, overweight or obesity, among others, and there is a lot of scientific evidence that shows the beneficial effects associated with physical exercise in the prevention and treatment of this type of disorders (Bull et al. 2017).

This study has had two major limitations: (1) the subject of study is very little developed in the scientific literature, although fortunately more attention is being paid to the field of physical exercise with virtual reality in health, since it is a high cost-effective intervention, and in times like the present (2020) in which the interventions with patients must be as safe as possible (with reference to the current pandemic context), it is an option that interests to be studied further. And (2) the low quality of the referenced evidence.

This leads us to propose the future research lines in which the subsequent studies will be developed. Future research should focus on more personalized interventions, addressing the specific characteristics of each type of disorder that exists within severe mental disorders, and the specific characteristics of each patient. The studies should protocolize an intervention that would be the standard, which would be used as the basis for building individual programs. Finally, it has been studied that the effects of these interventions are much greater if they are maintained over time, that is, if there is adherence to the practice. Therefore, strategies should be developed to promote this key aspect to improve the general health levels of these patients.

5 Conclusion

In conclusion, physical activity based on exergames or virtual reality could have positive effects on quality of life and depressive symptoms in people diagnosed with a severe mental disorder. However, this statement should be treated with caution, since there is a wide variety of mental illnesses where the effects caused by this type of interventions are very diverse. In addition, due to the heterogeneity presented in the study population, it is necessary to consider the personal characteristics of each patient to objectively assess the effects of this alternative therapy.

Clearly, more research is needed in the field of alternative therapies related to physical activity and mental disorders, especially those based on games or virtual reality. It is necessary the research of different types of exercises with different training parameters to find possible programs which can be used as treatments and could be translated into guidelines and recommendations for this population. So, we could see the application of physical activity in any of its modalities (e.g., exergames or virtual reality) in the clinical practice as part of the management and treatment of the severe mental disorders.

References

Bull, F., et al.: Physical activity for the prevention of cardiometabolic disease. In: Prabhakaran, D., et al. (Eds.), Cardiovascular, Respiratory, and Related Disorders, pp. 79–100. EEUU: The International Bank for Reconstruction and Development/The World Bank, Washington (DC) (2017)

Cohen, J.: Statistical Power Analysis for the Behavioral Sciences. Academic Press, New York, USA (2013)

Deenik, J., et al.: Physical activity and quality of life in long-term hospitalized patients with severe mental illness: a cross-sectional study. BMC Psych. **17**(1), 298 (2017). https://doi.org/10.1186/s12888-017-1466-0

Dellazizzo, L., et al.: Evidence on virtual reality-based therapies for psychiatric disorders: meta-review of meta-analyses. J. Med. Internet. Res. **22**(8), e20889 (2020). https://doi.org/10.2196/20889

Firth, J., et al.: Motivating factors and barriers towards exercise in severe mental illness: a systematic review and meta-analysis. Psychol. Med. **46**(14), 2869–2881 (2016). https://doi.org/10.1017/S0033291716001732

Fodor, L.A., et al.: The effectiveness of virtual reality based interventions for symptoms of anxiety and depression: a meta-analysis. Sci. Rep. **8**(1), 10323 (2018). https://doi.org/10.1038/s41598-018-28113-6

Huhn, M., et al.: Efficacy of pharmacotherapy and psychotherapy for adult psychiatric disorders: a systematic overview of meta-analyses. JAMA Psych. **71**(6), 706–715 (2014). https://doi.org/10.1001/jamapsychiatry.2014.112

James, S.L., et al.: Global, regional, and national incidence, prevalence, and years lived with disability for 354 diseases and injuries for 195 countries and territories, 1990–2017: a systematic analysis for the global burden of disease study 2017. Lancet **392**, 1789–1858 (2018). https://doi.org/10.1016/S0140-6736(18)32279-7

Maples-Keller, J.L., Bunnell, B.E., Kim, S.-J., Rothbaum, B.O.: The use of virtual reality technology in the treatment of anxiety and other psychiatric disorders. Harvard Rev. Psych. **25**(3), 103–113 (2017). https://doi.org/10.1097/HRP.0000000000000138

Piepmeier, A., Etnier, J.: Brain-derived neurotrophic factor (BDNF) as a potential mechanism of the effects of acute exercise on cognitive performance. J. Sport Health Sci. **4**(1), 14–23 (2015). https://doi.org/10.1016/j.jshs.2014.11.001

Romito-Di Giacomo, R.R., Menegay, H., Cicero, S.A., Herrup, K.: Effects of Alzheimer's disease on different cortical layers: the role of intrinsic differences in AB susceptibility. J. Neurosci. Official J. Soc. Neurosci. **27**(32), 8496–8504 (2007). https://doi.org/10.1523/JNEUROSCI.1008-07.2007

Rus-Calafell, M., et al.: Virtual reality in the assessment and treatment of psychosis: a systematic review of its utility, acceptability and effectiveness. Psychol. Med. **48**(3), 362–391 (2018). https://doi.org/10.1017/S0033291717001945

Swinnen, N., Vandenbulcke, M., Vancampfort, D.: Exergames in people with major neurocognitive disorder: a systematic review. Disabil. Rehabil. Assist. Technol. 1–14 (2020). https://doi.org/10.1080/17483107.2020.1785566

Välimäki, M., et al.: Virtual reality for treatment compliance for people with serious mental illness. Cochrane Database Syst. Rev. **8**(10), CD009928 (2014). https://doi.org/10.1002/14651858.CD009928.pub2

Vancampfort, D., et al.: Sedentary behavior and physical activity levels in people with schizophrenia, bipolar disorder and major depressive disorder: a global systematic review and meta-analysis. World Psych. Official J. World Psych. Assoc. **16**(3), 308–315 (2017). https://doi.org/10.1002/wps.20458

Wang, P.S., et al.: Use of mental health services for anxiety, mood, and substance disorders in 17 countries in the WHO world mental health surveys. Lancet **370**(9590), 841–850 (2007). https://doi.org/10.1016/S0140-6736(07)61414-7

Weinstein, G., et al.: Serum brain-derived neurotrophic factor and the risk for dementia. JAMA Neurol. **71**(1), 55–61 (2014). https://doi.org/10.1001/jamaneurol.2013.4781

Estimating Energy Expenditure During Active Virtual Reality Gaming

Julissa Ortiz-Delatorre[1], Ryan P. Durk[1,2], Alexandra Esparza[1], Adriana Ruiz[1], Maria Jericka Doaz[1], Marialice Kern[1,2], and James R. Bagley[1,2,3](✉)

[1] Exercise Physiology Laboratory, Department of Kinesiology, San Francisco State University, San Francisco, CA, USA
jrbagley@sfsu.edu
[2] Virtual Reality Institute of Health and Exercise, San Francisco, CA, USA
[3] Healthy Living for Pandemic Event Protection (HL-PIVOT) Network, San Francisco, CA, USA

Abstract. INTRODUCTION: Active virtual reality gaming (AVRG) provides an alternative option to traditional exercise modalities. Recent studies have shown that exercising using AVRG can lead to exercise intensities that meet guidelines for health benefits. Tracking caloric energy expenditure (EE) during exercise has shown to be effective for improving body composition and health, but no study to date has investigated differences between the accuracy of EE estimations from various devices during AVRG. OBJECTIVE: The purpose of this study was to compare two methods of tracking EE (movement displacement and heart rate) versus the laboratory "gold standard" (indirect calorimetry) during AVRG. METHODS: Eleven participants (6 females, 5 males, 22.5 ± 2.5 y) played 15-min of AVRG, while undergoing three different methods of EE tracking simultaneously: indirect calorimetry measuring oxygen consumption (VO_2), movement displacement data from the VR system (MOV), and a heart rate monitor (HR). RESULTS: Total estimated EE during the 15-min gameplay was 69.21 ± 10.85, 77.42 ± 24.63, and 134.00 ± 32.98 kcals for VO_2, MOV, and HR, respectively. The HR measure over-estimated kcals ($p < 0.0001$) compared to VO_2, but there was no significant difference between VO_2 and MOV EE estimates. CONCLUSION: These results suggest 1) movement displacement may be an accurate method to track EE while playing VR games compared to indirect calorimetry, and 2) new equations must be developed specifically for AVRG to improve HR estimates of EE during gameplay.

Keywords: Caloric expenditure · Movement displacement · Accelerometry · Heart rate · VO_2

1 Introduction

Over two thirds of the population is classified as overweight or obese in the United States [1]. Obesity and sedentary behaviors are associated with an increased risk for developing a variety of chronic diseases [2]. The American College of Sports Medicine (ACSM) and Center for Disease Control and Prevention (CDC) currently recommend ≥ 150 min

B. Sañudo Corrales and J. García-Fernández (Eds.): Tapasconference 2020, LNBE, pp. 66–73, 2022.
https://doi.org/10.1007/978-3-030-92897-1_8

of moderate-intensity aerobic activity every week [3, 4], but unfortunately only 21% of adults follow these guidelines [4]. This increase in sedentary behavior has partially attributed to the rise in amount of screen time (e.g., television, computers, and video games). Further, the global COVID-19 pandemic has reportedly increased screen time in some Western countries [5]. Obesity is a precursor to numerous diseases, but thankfully, it is a health condition can be prevented and potentially reversed by behavioral choices.

Research has shown that self-monitoring can promote successful weight control [6, 7]. Tracking caloric energy expenditure (EE) has been shown to be an effective way to monitor exercise and improve one's body composition [8]. A caloric deficit is needed to achieve weight loss, and by tracking caloric intake and EE, individuals can ensure they stay in a caloric deficit. In order for individuals to manage their weight and prevent obesity, they should have access to accurate applications and equipment that can aide them in the process of losing body fat to improve health.

While increased screen time is a serious health concern as it is typically associated with higher levels of sedentary behavior [9, 10], recent studies from our lab have demonstrated that "screen time" can actually be synonymous with physical activity [11]. There is a growing body of literature showing that active virtual reality gaming (AVRG) can be utilized as a valid exercise modality [12, 13]. Previous studies have looked at the effectiveness of AVRG as a form of physical activity as well as the effectiveness of playing these "exercise games" (termed exer-games) in various settings. The results showed that AVRG can serve as an effective alternative to traditional exercise methods and that the setting in which these games are played does not affect how enjoyable they are or the intensity they are played at [11, 14]. Due to these findings, the purpose of this study was to compare two commercially available methods of tracking EE (movement displacement and heart rate) versus the laboratory "gold standard" (indirect calorimetry) during AVRG.

2 Materials and Methods

Eleven volunteers (6 females, age = 22.5 ± 1.5 y; 5 males, age = 22.0 ± 1.6 y) were recruited from a large public university in California, USA. All participants were considered healthy and were asked to refrain from any physical activity the morning of testing. This study was reviewed and approved by the University's Institutional Review Board for Human Subjects and informed consent was acquired prior to any testing.

All participants reported to the laboratory for one testing session which consisted of playing one, 15-min game while tracking EE using three different modalities: oxygen consumption via indirect calorimetry (VO_2), movement displacement data from the VR system (MOV), and a heart rate monitor (HR). Upon arrival at the laboratory, demographic information was obtained (see Table 1) and anthropometric measures were recorded. Age, height, and weight were inputted into the corresponding devices prior to the start of the game for kilocalories (kcals) to be obtained. Participants were given a tutorial on how to perform the activities in the game via diagrams and verbal instructions. After becoming familiar with the game objective, participants were fitted with VO_2 facemasks and HR straps. Resting data was collected for 2 min prior to starting the 5-min familiarization period where participants warmed up. Participants continued to play the game for 15-min of data collection where all three methods simultaneous tracked EE.

Table 1. Participant demographics

	Age (y)	Height (m)	Weight (kg)	BMI (kg/m^2)
Female (n = 6)	22.5 ± 1.5	1.63 ± 0.06	65.91 ± 11.32	24.73 ± 3.25
Male (n = 5)	22.0 ± 1.6	1.77 ± 0.03	81.00 ± 15.69	25.95 ± 5.09

BMI, Body Mass Index. *Data are reported as mean ± SD.*

2.1 Energy Expenditure

Indirect Calorimetry (VO_2). Indirect calorimetry (TrueOne 2400: ParvoMedics Inc., Salt Lake City, UT) was unitized as the "gold standard" to estimate EE measuring oxygen consumption and carbon dioxide production [15].

Movement Displacement (MOV). Software (YUR.fit: YUR Inc., San Mateo, CA.) was downloaded on a VR headset (Oculus Quest: Facebook Inc., Menlo Park, CA.) that tracked the displacement of both the hand-held controllers as well as the headset. This software then used these measures as well as other anthropometric data to estimate EE during an AVRG (Note: this method is specific to AVRG, and is based on proprietary calculations from YUR Inc.).

Heart Rate Monitor (HR). Heart rate was monitored via a chest strap chest (Polar H10: Polar Electro Inc., Bethpage, NY) and tracked participants heart rate throughout a period of physical activity. The heart rate monitor was synced to the corresponding application (Polar beat: Polar Electro Inc., Bethpage, NY) which uses an algorithm to calculate estimated EE during physical activity based on the individual's heart rate and anthropometric measures (Note: this method is not specific to AVRG, and is based on proprietary calculations from Polar Electro Inc.).

2.2 Statistical Analyses

Statistical analyses were performed with SAS (Model 9.4 for Windows; Cary, NC). Descriptive statistics were run on all variables and displayed as mean ± standard deviation where possible. An ANOVA was utilized to determine if there was a significant difference between EE methodologies and if there was an interaction between sex or body mass index (BMI). A Tukey post hoc analysis was performed to confirm where any significant differences existed. This process was repeated for kcals and METs, and Bland-Altman analyses were run to assess agreement between VO_2 and MOV as well as VO_2 and HR. Finally, Fit Plot Differences were run between VO_2 and MOV as well as VO_2 and HR.

3 Results

Total estimated EE during the 15-min gameplay was 69.21 ± 10.85, 77.42 ± 24.63, and 134.00 ± 32.98 kcals for VO_2, MOV, and HR, respectively (Fig. 1). There was a significant effect for measurement method on the average total kcals used during exercise [F (2,30) = 22.69, $p < 0.0001$, $\eta 2 = 0.60$]. There were no significant differences between VO_2 and MOV; however, HR was statistically significant from VO_2 [t (30) = 6.17, $p < 0.0001$] and MOV [t (30) = 5.40, $p < 0.0001$], resulting in over-estimated kcals compared to VO_2. There was no significant interaction between sex or BMI on testing method.

There was a significant effect for device on the average METs used during exercise [F (2,30) = 13.44, $p < 0.0001$, $\eta 2 = 0.47$] (Fig. 1). There were no significant differences between VO_2 and MOV; however, HR was statistically significant from VO_2 [t (30) = 4.23, $p < 0.001$] and HR [t (30) = 4.71, $p < 0.0001$].

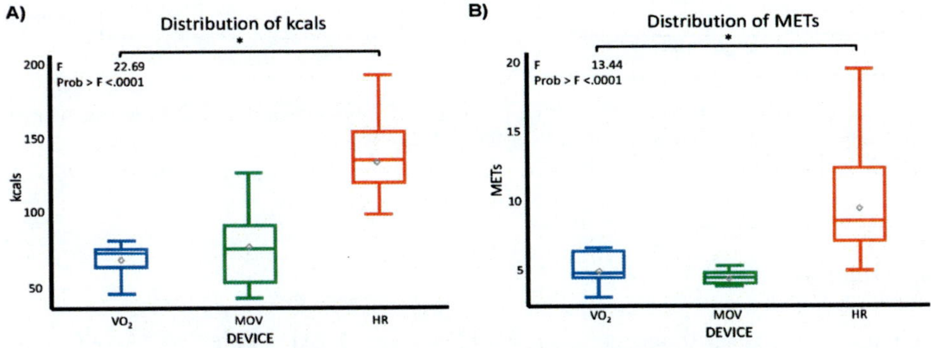

Fig. 1. Distribution of Calories (kcals) and metabolic equivalent (MET) estimations by device. A) Comparison of distribution of Kcals between VO_2, MOV, and HR. B) Comparison of distribution of METs between VO_2, MOV, and HR (*= $p < 0.05$).

Bland-Altman analyses report the mean bias (average difference between measures) and limits of agreement (LoA; 95% confidence intervals) between a criterion measure (VO_2) and other measures (MOV and HR). For VO_2 and MOV, the mean bias was –8.21 kcal and LoA was –64.75 to 48.33 kcal, indicating that MOV consistently over-estimated EE on average by 8.21 kcal. For VO_2 and HR, the mean bias was –64.75 kcal and LoA was –130.34 to 0.68 kcals, indicating that HR consistently over-estimated EE on average by 64.75 kcal (Fig. 2).

Fit Plot differences were conducted to determine if the variance was consistent among participants (homoscedastic = similar mean differences throughout the readings; heteroscedastic = discrepancy among mean differences throughout the readings). Fit Plot differences for VO_2 and MOV showed a significant model effect [F(1,9) = 53.21, $p < 0.001$, $R2 = 0.8556$]. Similarly, there was a significant model effect [F(1,9) = 75.93, $p < 0.001$, $R2 = 0.8823$] when looking at VO_2 and HR. These findings suggest the data are heteroscedastic and show more variability between EE devices at higher kcals.

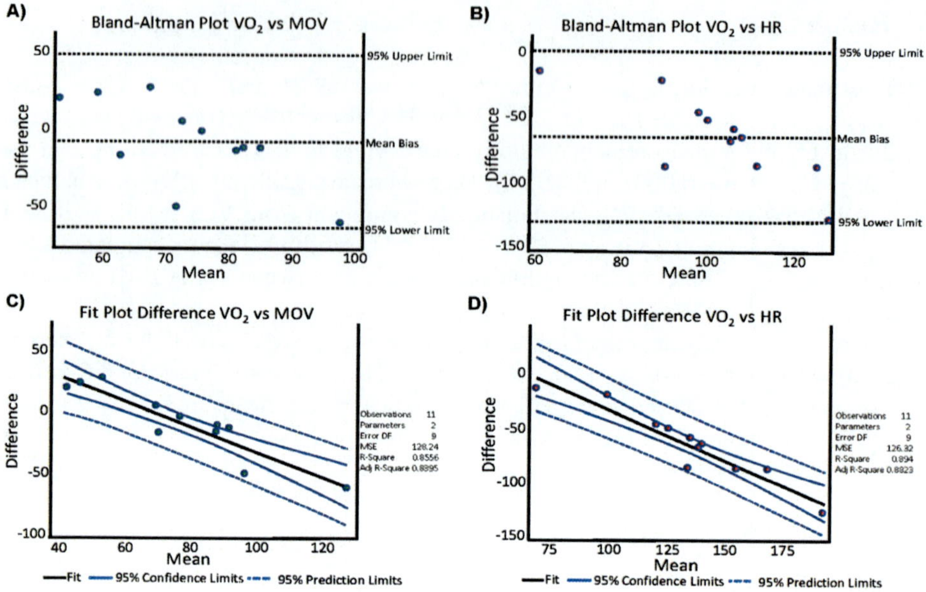

Fig. 2. Comparison of energy expenditure (EE) agreement between devices. A) Bland-Altman plot comparing VO_2 (criterion measure) vs. MOV. B) A) Bland-Altman plot comparing VO_2 vs. HR. C) Fit Plot difference comparing VO_2 vs. MOV. D) Fit Plot difference comparing VO_2 vs. HR.

4 Discussion

This study aimed to examine the accuracy of two methods of tracking EE while engaging in AVRG compared to a laboratory gold standard. The results suggest that 1) MOV is an accurate methodology to estimate EE and METs while 2) HR using a generic equation is an inaccurate method of tracking EE and METs in comparison to VO_2.

Previous studies have shown that heart rate can produce accurate measurements for EE after factors such as fitness level and age were accounted for [16, 17]. Variables such as fitness level, age, and sex have been shown to have an effect on the heartrate-O_2 relationship; therefore, if they are not taken into account, it could affect EE estimations [18]. In the present study, fitness level was not a factor that was taken into account by the HR device. This could be an explanation as to why HR values significantly overestimated EE. In order to improve the accuracy of EE estimations during AVRG, future research should focus on improving heart rate algorithms specifically for VR play.

The results suggest that MOV may be an accurate EE estimation method while engaging in AVRG since there were no significant differences between VO_2 and MOV. Although there is little research on the accuracy of movement displacement along with biometric information (e.g., height, weight, age) to estimate EE, research has been done on similar forms of activity tracking wearables (such as wrist-worn devices). These studies have consistently shown that activity trackers are typically an unreliable method of estimating EE [19–21]. In a meta-analysis by O'Driscoll et al. [20], the authors found that activity monitors typically underestimated EE and stated that they would be hesitant to consider these devices as adequately accurate. A possible explanation for the conflicting results between MOV and the devices used in previous studies could be that the MOV software in this study takes three points of movement displacement (headset, two controllers) into account while calculating EE estimation. Meanwhile, the devices used in previous research only take into account a single limb's movement. It is also important to note that the MOV in this study used movement displacement (quantifying movement in the x-y-z axis), as opposed to other devices which use accelerometry (instead of displacement) to assess EE.

Bland-Altman plots were run during data analyses to determine the agreement between two devices: VO_2 vs MOV and VO_2 vs HR. This indicated that EE estimations of both MOV and HR were consistently higher than the EE measured with VO_2; however, it is important to note that the differences in EE estimation between VO_2 and HR were much higher (average of 64.83 kcals) than those between VO_2 and MOV (average of 8.21 kcals). This can be problematic for AVRG players that use this platform as a form of exercise for extended periods of time as this can result in an extreme difference in EE estimation the longer the game is played. This could interfere with a player's health and fitness progression. AVRG developers should consider the validity of their EE estimates when designing games to improve metabolic fitness.

5 Limitations

Limitations to this study were centered around sample size, population, and VR game selection. The sample size of this study was relatively small ($n = 11$) which altered the model specification analysis of the Bland-Altman plot. The small sample size was enough to notice a trend in the Fit Plot difference analysis; however, there was no systematic relationship that could be established. The population in our study consisted of young (22.5 ± 2.5 y), healthy adults which is not representative of other groups in the general population. Additionally, this study used only one VR game to assess the accuracy of each device. The literature has suggested the accuracy of some activity trackers varies depending on activity type [20]. In this study, HR was not specifically designed for AVRG. Therefore, future studies should examine the accuracy of different methods of EE estimation (using VR specific algorithms), while engaging in different types of AVRG.

6 Conclusion

This study investigated the accuracy of two methods for tracking EE in comparison to a gold standard while engaging in AVRG. The data revealed significant differences in EE estimation between VO_2 and HR, but there were no significant differences between VO_2 and MOV. These results suggest 1) MOV may be an accurate method to track EE while playing VR games compared to indirect calorimetry, and 2) new algorithms must be developed specifically for AVRG to improve EE estimates during gameplay.

Acknowledgements. The authors would like to thank the participants for donating their time and Student Researchers (KIN 697/98 Senior Research; Spring 2020) for assisting with data collection.

This study was funded by a Joint Venture Grant (CSUPERB) to J.R. Bagley and the Virtual Reality Institute for Health and Exercise.

References

1. Center for Disease Control and Prevention: Adult Obesity Facts (2020). https://www.cdc.gov/obesity/data/adult.html. 26 Aug 2020
2. Rhodes, R., Mark, R., Temmel, C.: Adult Sedentary Behavior: A Systematic Review (2012). https://www.sciencedirect.com/science/article/pii/S074937971100910X. 26 Aug 2020
3. Trending Topic: Physical Activity Guidelines (n.d.). https://www.acsm.org/read-research/trending-topics-resource-pages/physical-activity-guidelines. 26 Aug 2020
4. American College of Sports Medicine: ACSM's Guidelines for Exercise Testing and Prescription. 10 Lippincott Williams & Wilkins, Philadelphia, PA (2017)
5. Colley, R.C., Bushnik, T., Langlois, K.: Exercise and screen time during the COVID-19 pandemic. Statist. Canada. **31**, 3–11 (2020). https://doi.org/10.25318/82003-x202000600001-eng
6. Baker, R.C., Kirschenbaum, D.S.: Self-monitoring may be necessary for successful weight control. Behav. Ther. **24**(3), 377–394 (1993). https://doi.org/10.1016/s0005-7894(05)80212-6
7. Burke, L.E., et al.: The effect of electronic self-monitoring on weight loss and dietary intake: a randomized behavioral weight loss trial. Obesity **19**(2), 338–344 (2010). https://doi.org/10.1038/oby.2010.208
8. Hartmann-Boyce, J., Johns, D.J., Jebb, S., Aveyard, P.: Effect of behavioural techniques and delivery mode on effectiveness of weight management. Systematic review, meta-analysis and meta-regression. Appetite **87**, 387 (2015). https://doi.org/10.1016/j.appet.2014.12.150
9. Hu, F.B.: Television watching and other sedentary behaviors in relation to risk of obesity and type 2 diabetes mellitus in women. JAMA **289**(14), 1785 (2003). https://doi.org/10.1001/jama.289.14.1785
10. Cleland, V.J., Schmidt, M.D., Dwyer, T., Venn, A.J.: Television viewing and abdominal obesity in young adults: is the association mediated by food and beverage consumption during viewing time or reduced leisure-time physical activity? Am. J. Clin. Nutr. **87**(5), 1148–1155 (2008). https://doi.org/10.1093/ajcn/87.5.1148
11. Gomez, D.H., Bolter, N., Lee, C.M., Bagley, J.R., Kern, M.: Metabolic cost and exercise intensity during active virtual reality gaming. Med. Sci. Sports Exerc. **50**(5S), 135 (2018). https://doi.org/10.1249/01.mss.0000535531.90524.97
12. Mcdonough, D.J., Pope, Z.C., Zeng, N., Liu, W., Gao, Z.: Comparison of college students' blood pressure, perceived exertion, and psychosocial outcomes during virtual reality, exergaming, and traditional exercise: an exploratory study. Games Health J. **9**(4), 290–296 (2020). https://doi.org/10.1089/g4h.2019.0196
13. Bond, S., Laddu, D.R., Ozemek, C., Lavie, C.J., Arena, R.: Exergaming and virtual reality for health: implications for cardiac rehabilitation. Curr. Probl. Cardiol. **46**, 100472 (2019). https://doi.org/10.1016/j.cpcardiol.2019.100472
14. Effect of Exercise Setting On Energy Expenditure And Enjoyment During Active Virtual Reality Gaming: 210 Board# 26 May 27 9: 30 AM-11: 00 AM
15. Crouter, S.E., Antczak, A., Hudak, J.R., Dellavalle, D.M., Haas, J.D.: Accuracy and reliability of the ParvoMedics TrueOne 2400 and MedGraphics VO2000 metabolic systems. Eur. J. Appl. Physiol. **98**(2), 139–151 (2006). https://doi.org/10.1007/s00421-006-0255-0
16. Strath, S.J., et al.: Evaluation of heart rate as a method for assessing moderate intensity physical activity. Med. Sci. Sports. Exer. **32**(Suppl.), 470–475 (2000). https://doi.org/10.1097/00005768-200009001-00005

17. Keytel, L., et al.: Prediction of energy expenditure from heart rate monitoring during submaximal exercise. J. Sports Sci. **23**(3), 289–297 (2005). https://doi.org/10.1080/02640410470001730089
18. Davis, J.A., Convertino, V.A.: A comparison of heart rate methods for predicting endurance training intensity. Med. Sci. Sports. Exer. **7**(4), 295–298 (1975). https://doi.org/10.1249/00005768-197500740-00010
19. Evenson, K.R., Goto, M.M., Furberg, R.D.: Systematic review of the validity and reliability of consumer-wearable activity trackers. Int. J. Behav. Nutr. Phys. Activi. **12**(1), 159 (2015). https://doi.org/10.1186/s12966-015-0314-1
20. O'Driscoll, R., et al.: How well do activity monitors estimate energy expenditure? A systematic review and meta-analysis of the validity of current technologies. Proc. Nutr. Soc. **77**(OCE4), E147 (2018). https://doi.org/10.1017/S0029665118001532
21. Shcherbina, A., et al.: Accuracy in wrist-worn, sensor-based measurements of heart rate and energy expenditure in a diverse cohort. J. Personal. Med. **7**(2), 3 (2017). https://doi.org/10.3390/jpm7020003

Digital Tools for Adapting Corporate Wellness Programmes to the New Situation Caused by COVID-19: A Case Study

José M. Núñez-Sánchez[1](✉), Ramón Gómez-Chacón[2], and Carmen Jambrino-Maldonado[3]

[1] Business Management Department, University of Málaga Doctoral Program in Economic and Business, Faculty of Economics, University of Málaga, 29013 Málaga, Spain
josemanuel.nunez@uma.es

[2] Department of Physical Education, CEU Cardenal Spínola, 41930 Sevilla, Spain

[3] Business Management Department, Faculty of Economics, University of Málaga, 29013 Málaga, Spain

Abstract. The COVID-19 pandemic has changed our lives completely: social distancing, confinement, telework and above all, negative consequences on our mental and physical health caused mainly by increased physical inactivity and sedentarism. Up to 34% of employees were doing telework during lockdown in Spain, and 30% remained working remotely even afterwards. Companies have developed corporate wellness programs that have proven its positive results before the pandemic. The problem with the situation caused by this pandemic is that workers cannot exercise anymore in the workplace facilities, so if they want to take care of their health and well-being at home, they should go digital. The aim of this research is to show how a well-known Spanish fitness company, O2 Centro Wellness, has applied all its digital knowledge to help its partner company, Mahou-San Miguel, to adapt its wellness program into the new reality, developing a successful online corporate wellness program using a fitness app and other social media tools. This case study may help other companies around the world as telework is growing and corporate wellness programs should adapt to this new reality.

Keywords: Healthy workers · Fitness App · Telework · COVID-19 · Corporate well-being

1 Introduction

The current COVID-19 coronavirus pandemic was declared by the World Health Organization (WHO) on March 11, 2020 [1]. This global pandemic coincides in time with another pandemic, the one, that was declared, also by the WHO in 2012 due to physical inactivity and sedentary lifestyle. However, this pandemic is very different in nature from that of COVID-19 since we have been living with it for several years [2].

B. Sañudo Corrales and J. García-Fernández (Eds.): Tapasconference 2020, LNBE, pp. 74–87, 2022.
https://doi.org/10.1007/978-3-030-92897-1_9

In addition, both pandemics influence each other, that is, physical inactivity affects the immune system, and therefore it would be more exposed to COVID-19, and, the confinement caused by COVID-19, causes an increase in sedentary lifestyle and a decrease of physical activity. The negative consequences of the lockdown caused by COVID-19 were reflected in a study carried out in different countries that highlighted the impact on mental, emotional, and even behavioural health [3]. In this sense, and being aware of these two major health problems, the WHO [4–10] has been publishing various documents and recommendations in order to mitigate these negative effects of both pandemics.

It seems clear that maintaining regular physical activity and routinely exercising in a safe home environment is an important strategy for healthy living during the coronavirus crisis. Home stay is a fundamental safety step that can limit infections from spreading them widely. But prolonged home stay can increase certain behaviour that leads to inactivity and contributes to anxiety and depression, which, in turn can lead to a sedentary lifestyle known to result in a range of chronic health conditions [11].

A study was carried out on the articles published about the benefits of regular physical activity preventing almost all the health problems that are associated with COVID-19. Exercising regularly becomes, therefore, the best natural medicine and the best way to prevent health complications from COVID-19 as shown in the next Table 1:

Table 1. 10 benefits of regular physical activity for preventing health problems that can be caused by Covid.

Benefits	Evidence	Author
Immune system	The immune system is very responsive to exercise	Nieman, DC and Wentz, L.M.; 2019 [12]
Lower incidence of respiratory infections	The practice of moderate or intense exercise is related to a lower incidence, duration, and severity of respiratory system infections	Grande et al.; 2020 [13]
Reduce risk of disease	There is a clear inverse relationship between moderate exercise training and illness risk	Nieman, DC and Wentz, L.M.; 2019 [12]
Lower incidence of mortality	Several studies suggest that doing physical exercise is associated with a lower incidence of mortality associated with influenza and pneumonia	Wong et al.; 2008 [14]

(continued)

Table 1. (*continued*)

Benefits	Evidence	Author
Reduces symptoms of depression	Exercise is moderately more effective than a controlled intervention to reduce symptoms of depression	Cooney et al.; 2013 [15]
Reduces anxiety	Aerobic exercise is an effective treatment for patients with increasing anxiety diagnosed in primary care, as well as with anxiety disorders	Aylett et al.; 2018 [16]
Better mood and reduces alcohol consumption	The absence or decrease of physical activity increases the risk of weight gain due to not spending energy. In addition, anxiety, depression, mood and boredom could be offset, at least I n part, by increased food and alcohol consumption	The Foundation for Alcohol Research and Education, 2020 [17]
Stress reduction and better sleep quality	Among the recommendations for managing sleep problems in this period, by the working group of the CBT-II Academy (cognitive behavioral therapy for insomnia, we will find the practice of regular exercise in daylight	Altena et al.;2020 [18]
Reduces severity of infections	Among the benefits described, in physical active people, for both the immune and respiratory systems, we can find the reduction of the severity of infectious diseases	King et al.;2019 [19]
Anti-inflammatory properties	Exercise training has anti-inflammatory properties	Nieman, DC and Wentz, L.M; 2019 [12]

1.1 Corporate Wellness

There is no doubt about it; the workplace is a conducive environment to promote physical activity, since companies achieve very positive results such as greater productivity and improving employees' health, all of which have an impact on physical and psychological benefits [20, 21]. There is ample evidence of the positive relationship between regular physical activity and increasing labor productivity [22], in addition to achieving decrease in absenteeism [23], increasing job satisfaction [24], and work capacity [25]. On the other hand, corporate wellness programmes impact on the employees' physical aspects, improving their muscle tone and general physical condition [26], reducing health risks [27], improving psychophysiological and social parameters related to the relief of musculoskeletal pain [28], decreasing overweight [29], improving the body mass index, body weight, fat mass, cholesterol, fitness and sleeping quality [30].

Corporate well-being programmes also have a positive impact on psychosocial aspects, such as emotional stability, optimism, enthusiasm, and psychological well-being [31], trust [32], greater perception of the ability to control memory [33], the willpower to work, the feeling of well-being and interpersonal relationships [28], and with an improvement in the resilience of employees [34]. In addition, Gil-Beltrán, Meneghel, Llorens and Salanova [35] showed that, in order to reduce stress levels, improve job satisfaction levels, and make work emotions more positive, physical activity must be performed regularly.

1.2 Telework

It is evident that, associated with the pandemic, teleworking has had an exponential development, and this is transforming the normality of working. Teleworking, defined as a way of working that is carried out in a location far from the headquarters, working offices or production facilities, separating the worker from personal contact with colleagues who are in the same office, and the new technology makes this separation possible by facilitating information and communication [36]. In this sense, it is worth highlighting the Adecco Group report, "Resetting Normal" [37]: carried out in June 2020, in which 8,000 workers between 18 and 60 years old, with contracts of more than 20 h, were interviewed in 8 countries. The main conclusions are that after the pandemic, 75% of employees want flexibility between office and remote work (hybrid model), in Spain, this percentage reaches 78%, the objective of maintaining physical health has been one of the 5 most important aspects for their working life after the pandemic, for 80% of the Spanish workers surveyed. The main conclusion of the study is that the future of work is Flexible: The world is ready for "hybrid working" [37].

For all the above, companies that take care of their workers' well-being should adapt their wellness programs at their workplaces to this new hybrid environment caused by COVID-19.

1.3 New Technologies

The only way to adapt to this new environment of confinements and telework is through new technologies [38]. It is known that there are positive relationships between e-lifestyles, perceived ease of use, perceived usefulness and attitude toward fitness apps and the importance of these e-lifestyles as predecessors of the use of fitness app [39]. These authors explain that e-lifestyles are individual forms of behavior in the digital environment that reflect the values, activities, interests, and opinions of consumers. Likewise, fitness apps are considered technological tools to promote physical activity online. In their research they concluded the importance of e-lifestyles with the intention of use fitness apps and therefore promoting physical activity through online fitness services. The inclusion of virtual fitness classes is leading to an increase in online offers available for consumers, expanding the hours that consumers can exercise and leading to a greater variability of customer services [40]. In addition, two studies identified the importance of social networks in the process of customer retention. They concluded that the social networking sites and the use of new technologies have positive influence on client retention in fitness clubs [41]. During confinement there was an increase in offering digital channels that help guided physical activity through the use of safe, simple and easy to implement programmes and applications that cover activities related to cross-fit, yoga or dance activities for the general population or for the improvement of physical and mental well-being of older adults [42].

Fitness plays an important role in Corporate Wellness; therefore, it seems appropriate to conclude that the use of social media and technology can help us with the retention and physical activity of workers in our corporate sports programmes. Taking it into account, the influence of e-lifestyle on the intention of using a fitness app, it seems advisable to add a Fitness app to all these digital tools so that everything works and, at the same time, maintain constant communication with the employees with all the digital tools at our disposal, for example video conferencing systems, chat, mobile, social networking sites, and even via App.

1.4 Corporate Wellness Programmes in the COVID-19 Scenario

The preliminary results of an international survey indicate a negative effect of home confinement on Physical Activity and diet behavior with a significant increase in sitting time and unhealthy diet, indicative of a more sedentary lifestyle [43]. Companies should adapt their corporate wellness programme.s to this new environment, using all the digital tools at their disposal to avoid these negative effects. Based on this exploratory research, the current scenario is described in the following Fig. 1:

1.5 Objective

The aim of this research, that took place from April to June 2021, is to show how it is possible, with the help of digital tools, to adapt in a short period of time, a corporate wellness programme, to the new environment caused by the COVID-19 confinement and its consequences on the expansion of teleworking. On the other hand, the present situation will cause a big change in the Corporate Wellness programme model that, thanks to new technologies, should adapt to the new scenario and become hybrid, with an important physical part but complemented with a virtual part too.

Therefore, this case study may help other companies around the world to adapt their corporate programmes to this new reality.

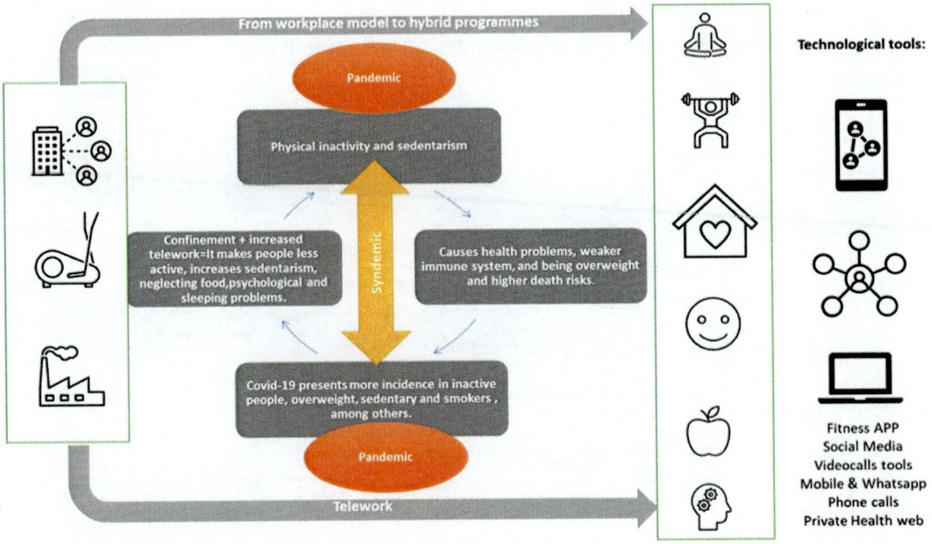

Fig. 1. Adaptation of corporate wellness programs to the new COVID 19 and telework scenario.

2 Methodology

The case study method was chosen to carry out a single study and analyse the wellness programme of Mahou-Miguel during the telework associated to the lockdown period, taking place from April to June 2020.

The case study method has been selected as it has been widely used in management research as a source of knowledge in various areas, as an example to follow, as a sample of what can happen, or as a source of experience [44].

Mahou-San Miguel is the leading Spanish company in the beer sector in Spain, with more than 34% market share, and a wide international presence in more than 70 countries in the world. As said in the company, they are a big family of more than 4,250 professionals spread over different locations and production centres. Their values are, the drive for innovation, believing in people, the search for excellence, enjoying doing things well and creating sustainable businesses and environments. It is a benchmark company in corporate wellness programmes in Spain. Its evolution and milestones can be observed in the following Fig. 2.:

The Mahou-San Miguel corporate wellness programme was launched in 2000, named Programa a tu Salud (At your Health). During this period, from 2000 to 2017, the programme won different national awards, as shown in Fig. 2. Of particular note was the Socio-economic assessment of a Physical Activity Programme for the employees of a company 2013 done in cooperation with the Consejo Superior de Deportes de España and Instituto Nacional de Seguridad e Higiene en el Trabajo, where it was demonstrated, for the first time in Spain, the programme's return on investment (R.O.I.) of 2.8 € per each euro invested [45]. In 2017, the company decided to evolve its programme and the Cuidarme Programme was born, highlighting the creation and opening of a corporate gym with a health area at the company's headquarters, for which they relied on the expert

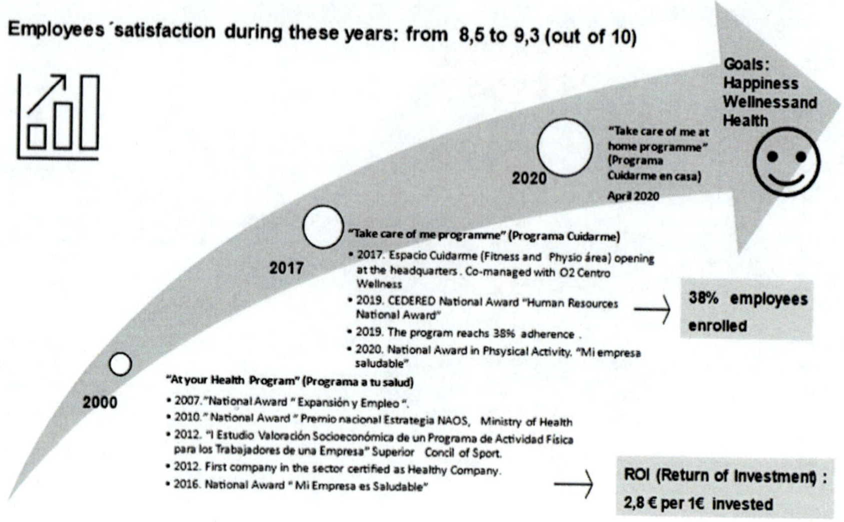

Fig. 2. The Mahou- San Miguel's corporate wellness program: Evolution and milestones.

management of O2 Centro Wellness, a company, that has accompanied them ever since. This is a company with large experience in the Fitness and Wellness Industry in Spain, offering therefore, another interesting point of view for this study.

The research was organised in 6 sequential stages as follows:

1. Review of literature
2. Analysis of the situation and documentation
3. Observation phase
4. Semi-structured interviews
5. Analysis of results
6. Discussion and Conclusions

The first step was a deep review of literature. Once it was completed, it continued with a thorough analysis of available documentation of this wellness programme and its trajectory from its birth to the present time, concentrating on the period that goes from the declaration of the State of Alarm and subsequent confinement, Government of Spain, March 14, 2020 [46] until the end of it on the 5th of June [47]. The research continued with direct observation to understand how the programme works in order to get to know in detail the particularities of the researcher's scenario and improve the validity of the study.

The study was completed with two semi-structured interviews (qualitative approach), the first one with the head of the project in Mahou-San Miguel; the Wellbeing Manager, who gave his vision as the responsible for the success of this corporate wellness programme. The second interview was with the O2 Centro Wellness business unit coordinator at Mahou-San Miguel. These two professionals have been working together on developing, managing, and executing the programme in the company's area of physical activity and health, set up at its headquarters (The Health Area). In this second interview, the objective was to find out how they have helped the brewing company to adapt its corporate wellness programme to the times of COVID-19.

3 Results

The results of this research, mainly obtained from the two semi-structured interviews conducted, are exposed in this section.

Firstly, the interview with Mahou-San Miguel Wellness Manager was conducted. It can be highlight that "Take care of me" (Cuidarme) is an innovative health and wellness programme integrated into the Mahou-San Miguel people development model that proposes to accompany them in the process of improving their own health. At the beginning of the interview, the focus was on the basic pillars of their wellness programme. This programme addresses all dimensions of well-being: physical, emotional, social and professional, underlining that "our actions are aimed at making all professionals feel safer, healthier and happier".

Second, it was asked how Covid has affects these dimensions. These are even more important in the new Covid reality, in which the company's proposal is to accompany all professionals in their new day-to-day life and help them to be competent, competitive, and balanced professionals. One of the main concerns that the company has had, and continues to have, when facing the current situation of the pandemic has been to preserve the health of professionals at every moment.

Third, it was important to know more about the digitalization of the company. This has been one of the fundamental elements of the programme in recent times, being considered a strategic pillar of the company and one of the clearest bets to improve the employee experience.

Last, enquiring about special measures adopted, was of special relevance. During this pandemic various organizational and strategic measures have been adopted in this company. The wellness programme was one of them, acquiring special importance in this period. All possible actions were taken to offer these programme at people's homes, virtual sessions or on-demand, mainly through the Cuidarme Web platform and the fitness-App. Before this, the main risks of lockdown were identified. Once the analysis was done, multiple actions were launched, taking advantage of the benefits of technology. Thanks to technology "teleworking is another opportunity in the development of this programme allowing us to foment their engagement to it, not only to the workers, but also to their families".

Secondly, the interview with the O2 Centro Wellness coordinator in the Mahou-San Miguel business unit was conducted.

The interview started enquiring what type of measures were taken when confinement started. Among the different strategies to take care of employees in this period there could be find direct communication with workers through the fitness application, emails, Microsoft Teams, phone and WhatsApp, updated weekly routines (Workouts Of the Week) uploaded to the app, with 3 different objectives and 3 levels, uploading fitness content to the corporate wellness private Mahou-San Miguel Website, a new YouTube private channel for broadcasting online activities, uploaded later on for its use on-demand

and finally customised training routings for employees, adapted to this confinement situation.

The next question was if the programme adopted in this period was successful and how they measure this. Online surveys were designed by O2 Centro Wellness for this purpose, and sent by Mahou-San Miguel to their employees by email. Finally, he added: "After analysing the results, it can be concluded that the programme has been very successful both for O2 Centro Wellness and for Mahou San Miguel".

When asked about the data supporting his sentence, the following figures were highlighted:

1. 60.3% of the employees were doing more sports or the same as before the pandemic.
2. 73,4% of the workers were satisfied of very satisfied with the programme
3. 501 training programmes were updated in this period
4. 46,1% of workers trained online with the O2CW team during this period. This percentage is 8% more than the percentage of workers who were following the programme before Covid-19.

To conclude the interview, enquiring about the technological tools more used by the employees for training in this period, was of special importance. The most used tool was the APP, followed by the private YouTube channel created for this occasion, and then the private Mahou-San Miguel Health website.

The "Take care of me at home" (Cuidarme en Casa), has taken advantage of all the existing technologies in the company that made it possible to implement teleworking some years ago as a work-life balance measure. Furthermore, these technologies, apart from others, have been used in the well-being programme, to improve customer satisfaction, maintaining constant and multichannel communication with the employees, not

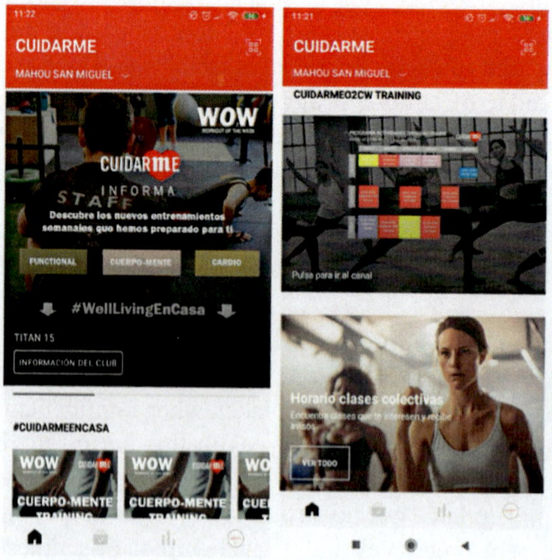

CONCEPT	RESULT
With APP Cuidarme downloaded	66%
% workers following the program	38%
More used training tool	54,8% APP Cuidarme
Youtube	30,1%
Updated training programs during COVID-19	501
Physycial activity similar or higher during COVID-19	60,3%
Satisfied or very satisfied with the program during Covid-19	73,4%

ONLINE ACTIVITIES	AVERAGE
Youtube online classes per week	12/week
Average atendance per class	3,5/class
Classes watched per month	707/month
Most demanded	Hiit Training, Pilates and Stretching

Fig. 3. Graphical overview of the data obtained in the interview with the O2 Centro Wellness coordinator.

only from the headquarters, but from all the locations in Spain and worldwide, and also with the sales team and, as highlighted by the Wellness Manager: "placing the person as the core of everything".

In Fig. 3, images of the Take care of me (Cuidarme) App can be observed, showing examples of weekly workouts and virtual activities, and data obtained from the questionnaires sent by Mahou-San Miguel.

4 Discussion and Conclusions

Technology and digital tools have helped many companies and workers in these difficult times so that they could continue with their work (telework) despite confinement and limitations in mobility. On the other hand, the confluence at the same time of two pandemics directly and seriously affect the employees' health, the COVID-19 pandemic and the pandemic of physical inactivity and sedentary lifestyle has generated various recommendations from the WHO concerned about the devastating effects that confinement can have on people's health [9].

It has also been confirmed the positive effects that corporate wellness programs have for the employees, and therefore, for the company, and the importance of new technologies and fitness app to improve the client retention (the employee).

Taking into account all these previous factors, the aim of the research has been achieved, showing how a well-known Spanish fitness company, O2 Centro Wellness, has applied all its digital knowledge to help its partner company, Mahou-San Miguel, to adapt its wellness program into the "new reality", developing a successful online corporate wellness programme using a fitness app (mywellness), communication tools (videocalls with Microsoft Teams, Emails, WhatsApp or Mobile), and social media tools (YouTube or Instagram) and a private corporate health website. It is important to highlight that 60.3% were doing sports as before Covid-19, or even more, which contrasts with research that concludes that the Spanish adult population decreased daily physical activity and increased sedentary time during confinement [48].

Another aim was to show other companies and institutions a practical case of how a leading company in corporate wellness program has been able, thanks to these new technologies, to continue taking care of the Mahou-San Miguel "family" successfully.

It can be inferred that from this moment on, all successful corporate wellness programmes should become hybrid programs, combining face-to-face and, also virtual. Taking advantage of all the digital tools at our disposal, highlighting the Fitness APP, social media and a health and wellness website. During confinement, it was possible to observe an increase in the offer of digital channels that help the guided practice of physical activity throughout the use of safe, simple and easy to implement programs and applications that cover activities related to cross-fit, yoga or dance activities for the general population or for the improvement of physical and mental well-being in older adults [42]. According to the findings in this successful case study, it can be concluded that it is highly recommended implementing these hybrid corporate well-being programmes, due to all the physical, mental and social benefits it brings to employees and, therefore to the company [20–35], even more in this COVID-19 times. This is in line with Núñez-Sánchez et al. (2021) [49], who concluded that physical inactivity and increased sedentary lifestyles triggered by Covid-19 and teleworking are having negative consequences for the physical and mental health of employees, which is why it is essential for companies that want to look after the health of their employees, while they work at home, to adapt their wellness programmes to the new situation as soon as possible.

This study has some limitations. Firstly, the company has more than 20 years of experience in the management of corporate wellbeing programmes. Secondly, it is a large company with more resources than smaller companies. Thirdly, they had already implemented a fitness app 4 years before the pandemic. These facts could limit the model's suitability for companies with different characteristics.

In conclusion, the findings of the current study suggest it is possible to successfully adapt worksite corporate wellness programs to the new digital environment caused by COVID-19, and the increase in telework that, according to different research [37], it will continue to grow in the future, helping to improve their employees' health and wellbeing. Corporate well-being programmes have shown to have a very positive impact on the organisations that implement them as well as on the health and well-being of their employees, therefore, the knowledge of proposals from a leading company in corporate well-being programmes, is of interest as they could be replicated in other companies in different sectors. [49].

Acknowledgements. We would like to thank Mahou-San Miguel and O2 Centro Wellness for their total collaboration to make this study possible.

References

1. World Health Organization (WHO). Coronavirus disease 2019 (COVID-19) Situation Report–51 (2020). https://www.who.int/docs/default-source/coronaviruse/situation-reports/20200311-sitrep-51-covid-19.pdf?sfvrsn=1ba62e57_10. Accessed 30 Jan 2021
2. Kohl, H.W., et al.: Lancet physical activity series working group. The pandemic of physical inactivity: global action for public health. The Lancet 2012, (London, England), **380**(9838), 294–305 (2012). https://doi.org/10.1016/S0140-6736(12)60898-8
3. Ammar, A., et al.: Effects of COVID-19 home confinement on eating behaviour and physical activity: results of the ECLB-COVID19 international online survey. Nutrients **12**(6), 1583 (2020)
4. World Health Organization (WHO): Global Strategy on Diet, Physical Activity and Health (2004). https://www.who.int/dietphysicalactivity/strategy/eb11344/strategy_english_web.pdf. Accessed 20 Nov 2020
5. World Health Organization: Una guía para enfoques basado en población para incrementar niveles de actividad física. Recuperado de (2008). https://www.who.int/dietphysicalactivity/pa/es/. Accessed 20 Nov 2020
6. World Health Organization. Global recommendations on physical activity for health, (2010). https://www.who.int/dietphysicalactivity/publications/9789241599979/en/. Accessed 20 Nov 2020
7. World Health Organization. (WHO). Global action plan for the prevention and control of noncommunicable diseases 2013–2020. 2013 https://apps.who.int/iris/bitstream/handle/10665/94384/9789241506236_eng.pdf;jsessionid=ADC589EBA2A1E6B5C20D2D6708041334?sequence=1. Accessed 20 Nov 2020

8. World Health Organization (WHO). Global action plan on physical activity 2018–2030: more active people for a healthier world (2018). https://apps.who.int/iris/bitstream/handle/10665/272722/9789241514187-eng.pdf?ua=1. Accessed 20 Nov 2020
9. World Health Organization. Stay physically active during self-quarantine (2019). http://www.euro.who.int/en/health-topics/healthemergencies/coronavirus-covid-19/novel-coronavirus2019-ncov-technical-guidance/stay-physically-activeduring-self-quarantine/_recache#article. Accessed 20 Nov 2020
10. World Health Organization: Campañas mundiales de salud pública de la OMS: #sanosencasa (2020). https://www.who.int/es/campaigns/connecting-the-world-to-combat-coronavirus/healthyathome. Accessed 20 Nov 2020
11. Chen, P., Mao, L., Nassis, G.P., Harmer, P., Ainsworth, B.E., Li, F.: Coronavirus disease (COVID-19): The need to maintain regular physical activity while taking precautions. J. Sport Health Sci. **9**(2), 103–104 (2020). https://doi.org/10.1016/j.jshs.2020.02.001
12. Nieman, D.C., Wentz, L.M.: The compelling link between physical activity and the body's defence system. J. Sport Health Sci. **8**(3), 201–217 (2019). https://doi.org/10.1016/j.jshs.2018.09.008
13. Grande, A.J., Keogh, J., Silva, V., Scott, A.M.: Exercise versus no exercise for the occurrence, severity, and duration of acute respiratory infections. Cochrane Database Syst. Rev. **4**(4), CD010596 (2020). https://doi.org/10.1002/14651858.cd010596.pub2
14. Wong C.M., et al.: Is exercise protective against influenza-associated mortality? PLoS One **3**(5), e2108 (2008). https://doi.org/10.1371/journal.pone.0002107
15. Cooney, G.M.: Exercise for depression. Cochrane Database Syst. Rev. **9**, CD004366 (2013). https://doi.org/10.1002/14651858.CD004366.pub6
16. Aylett, E., Small, N., Bower, P.: Exercise in the treatment of clinical anxiety in general practice–a systematic review and meta-analysis. BMC Health Serv. Res. **1**, 1–18 (2018)
17. The Foundation for Alcohol Research and Education. Alcohol sales & use during COVID-19. Deakin, ACT: FARE (2020). https://fare.org.au/wp-content/uploads/COVID-19-POLL.Pdf. Accessed 19 Apr 2020
18. Altena E, et al.: Dealing with sleep problems during home confinement due to the COVID-19 outbreak: practical recommendations from a task force of the European CBT-I Academy. J Sleep Res [Internet] (2020). https://www.ncbi.nlm.nih.gov/pubmed/32246787
19. King, A.C., et al.: Physical activity promotion: highlights from the 2018 physical activity guidelines advisory committee systematic review. Med. Sci. Sports. Exerc. **51**(6), 1340–1353 (2018). https://doi.org/10.1249/mss.0000000000001945
20. Gómez-Chacón, R., Grimaldi-Puyana, M., Bernal-García, A., Fernández-Gavira, J.: La práctica de actividad física y su relación con la satisfacción laboral en una organización de alimentación. J. Sports. Econ. Manage. **2**(6), 85–98 (2016)
21. Madden, S., et al.: The effect of workplace lifestyle programs on diet, physical activity and weight-related outcomes for working women: a systematic review. Obes. Rev. **21**, e13026 (2020). https://doi.org/10.1111/obr.13026
22. Whitsel, L.P., et al.: Assessing physical activity, sedentary behavior, and cardiorespiratory fitness in worksite health promotion. Am. J. Health Promot. **33**(2), 318–326 (2019). https://doi.org/10.1177/0890117118816750e
23. Van Amelsvoort, L., Spigt, M.G., Swaen, G.M.H., Kant, I.: Leisure time physical activity and sickness absenteeism; a prospective study. Occup. Med. **56**(3), 210–212 (2006). https://doi.org/10.1093/occmed/kqj025
24. Laroche, E., L'Espérance, S., Mosconi, E.: Use of social media platforms for promoting healthy employee lifestyles and occupational health and safety prevention: a systematic review. Safety Sci. **131**, 104931 (2020). https://doi.org/10.1016/j.ssci.2020.104930

25. Lusa, S., et al.: Interventions to promote work ability by increasing sedentary workers' physical activity at workplaces - a scoping review. Appl. Ergon. **2020**(82), 102962 (2020). https://doi.org/10.1016/j.apergo.2019.102961
26. Lock, M., Post, D., Dollman, J., Parfitt, G.: Efficacy of theory-informed workplace physical activity interventions: a systematic literature review with meta-analyses. Health Psychol. Rev. (2020). https://doi.org/10.1080/17437199.2020.1718527
27. Groenesteijn, L., Commissaris, D.A.C.M., Van Den Berg-Zwetsloot, M., Hiemstra-Van Mastrigt, S.: Effects of dynamic workstation Oxidesk on acceptance, physical activity, mental fitness and work performance. Work **54**(4), 773–778 (2016)
28. Silva, J.M.N., et al.: A worksite physical activity program and its association with biopsychosocial factors: an intervention study in a footwear factory. Int. J. Industr. Ergon. 6973–6979 (2019). https://doi.org/10.1016/j.ergon.2018.10.000
29. Aittasalo, M., et al.: Moving to business - changes in physical activity and sedentary behavior after multilevel intervention in small and medium-size workplaces. BMC Public Health **17**(1), 319 (2017). https://doi.org/10.1186/s12889-017-4229-4
30. Flahr, H., Brown, W.J., Kolbe-Alexander, T.L.: A systematic review of physical activity-based interventions in shift workers. Prev. Med. Reports **10**, 323–331 (2018). https://doi.org/10.1016/j.pmedr.2018.04.004
31. Boix-Vilella, S., León-Zarceño, E., Serrano-Rosa, M.Á.: Niveles de salud psicológica y laboral en practicantes de Pilates / Levels of Psychological and Occupational Health in Pilates Adherents. Revista Costarricense de Psicología. **37**(2), 145 (2018). https://doi.org/10.22544/rcps.v37i02.04
32. Díaz, X., Mena, C.B., Rebolledo, A.C.: Propuesta de un programa de promoción de la salud con actividad física en funcionarios públicos. Praxis Educativa. **15**(15), 104–109 (2012)
33. Loprinzi, P.D.: Association between habitual physical activity on episodic memory strategy use and memory controllability. Health. Promot. Perspec. **9**(1), 65–70 (2019). https://doi.org/10.15171/hpp.2019.07
34. Gerber, M., Jonsdottir, I.H., Lindwall, M., Ahlborg, G.: Physical activity in employees with differing occupational stress and mental health profiles: a latent profile analysis. Psychol. Sport. Exerc. **15**(6), 649–658 (2014). https://doi.org/10.1016/j.psychsport.2014.07.012
35. Gil-Beltrán, E., Meneghel, I., Llorens, S., Salanova, M.: Get vigorous with physical exercise and improve your well-being at work! Int. J. Environ. Res. Public Health **17**(17), 6384 (2020). https://doi.org/10.3390/ijerph17176384
36. Ramírez, S.L.C., Rúa, N.E.G.: The concept of telecommuting: aspects for health and safety in employment. 10. CES Salud Pública **5**, 82–91 (2014)
37. The Adecco Group. Resetting Normal: redefiniendo la nueva era de trabajo, 2020. https://www.adeccoinstitute.es/informes/resetting-normal-redefiniendo-la-nueva-era-del-trabajo/. Accessed 30 Jan 2021
38. Iglesias-Sánchez, P.P., Vaccaro Witt, G.F., Cabrera, F.E., Jambrino-Maldonado, C.: The contagion of sentiments during the COVID-19 pandemic crisis: the case of isolation in Spain. Int. J. Environ. Res. Public Health **17**, 5918 (2020)
39. García-Fernández, J., Gálvez-Ruiz, P., Grimaldi-Puyana, M., Angosto, S., Fernández-Gavira, J., Bohórquez, M.R.: The promotion of physical activity from digital services: influence of e-lifestyles on intention to use fitness apps. Int. J. Environ. Res. Public Health **17**(18), 6839 (2020)
40. Baena-Arroyo, M.J., García-Fernández, J., Gálvez-Ruiz, P., Grimaldi-Puyana, M.: Analyzing consumer loyalty through service experience and service convenience: differences between instructor fitness classes and virtual fitness classes. Sustainability **12**(3), 828 (2020)
41. Barbosa, H.F., García-Fernandez, J., Carrión, G.C.: Influência das tecnologias na retencao de socios em ginásios. Revisao Sistemática. Movimento **26**, 26070 (2020)

42. Angosto, S., García-Fernández, J., Valantine, I., Grimaldi-Puyana, M.: The intention to use fitness and physical activity apps: a systematic review. Sustainability **12**(16), 6641 (2020)
43. Ammar, A., et al.: Effects of COVID-19 home confinement on eating behaviour and physical activity results of the ECLB-COVID19 international online survey. Nutrients. 12(6), 15-83 (2020)
44. Mariotto, F.L., Zanni, P.P., Moraes, G.S.M.: What is the use of a single-case study in management research? Revista de Administração de Empresas **54**(4), 358–369 (2014). https://doi.org/10.1590/s0034-759020140402
45. Consejo Superior de Deportes de España and Instituto Nacional de Seguridad e Higiene en el Trabajo (INSHT). Socio-economic assessment of a Physical Activity Programme for the employees of a company (2013). https://www.insst.es/documents/94886/96076/valoracion+socio+economica/d1ab56dd-85bf-47cf-bba1-35d16244f839. Accesed 30 Jan 2021
46. de España, G.: Real Decreto 463/2020, de 14 de marzo, por el que se declara el estado de alarma para la gestión de la situación de crisis sanitaria ocasionada por el COVID-19 (2020). https://www.boe.es/buscar/pdf/2020/BOE-A-2020-3692-consolidado.pdf. Accessed 20 Nov 2020
47. de España, G.: Real Decreto 555/2020, de 5 de junio, por el que se prorroga el estado de alarma declarado por el Real Decreto 463/2020, de 14 de marzo, por el que se declara el estado de alarma para la gestión de la situación de crisis sanitaria ocasionada por el COVID-19 (2020). https://www.boe.es/buscar/pdf/2020/BOE-A-2020-5767-consolidado.pdf. Accessed 20 Nov 2020
48. Castañeda-Babarro, A, Arbillaga-Etxarri, A, Gutiérrez-Santamaria, B., Coca, A.: Impact of COVID-19 confinement on the time and intensity of physical activity in the Spanish population. Research Square (2020). https://doi.org/10.21203/rs.3.rs-26074/v0
49. José M. Núñez-Sánchez, Ramón Gómez-Chacón, Carmen Jambrino-Maldonado, Jerónimo García-Fernández,: Corporate Well-Being Programme in COVID-19 Times. The Mahou San Miguel Case Study. Sustainability **13**(11), 6189 (2021). https://doi.org/10.3390/su13116189

Analysis of the Sport-Related Habits of Fitness Centers Users According to Their Institutional Mobile Applications Downloads

Manel Valcarce[1](✉) and Salvador Angosto[2]

[1] Valencian International University (VIU), Valencia, Spain
manel.valcarce@campusviu.es

[2] Department of Physical Activity and Sport, Faculty of Sport Sciences, University of Murcia, Murcia, Spain
salvador.a.s@um.es

Abstract. The digitalization and incorporation of technology in the fitness sector and sports facilities is becoming an important element in the relationship of fitness centers with their users. Consequently, there is a need to know the influence of digitalization on the habits and behavior of the users, and compare those with the behavior of users who do not use the digital tools. This study aimed at analyzing the changes in sports habits of fitness centers users depending on whether they had downloaded the center's mobile application, or not. In order to do this, an analytical-descriptive study was carried out, with 1,285 participants (55% female and 44.8% male participants). A set of variables such as the use of the application and various sport habits (frequency, duration, level of fitness experience and activity practiced) were taken into account. Significant differences in users' sports habits were registered from the time of registration for the fitness center and the time of this study. Differences were also observed between those who downloaded the mobile application and those who did not. Those who used the application presented higher levels of sports activity, increasing their frequency of weekly attendance, spending more time when training and having a higher level of fitness experience than those who did not download the application. In addition, they did more group training and less swimming. A significant trend towards use of the fitness and bodybuilding room and training in the cardiovascular area was also observed. Future studies may assess the influence of technology and the use of mobile applications according to the business model of the fitness center, and depending on the customer profile (e.g. age, gender or socio-demographic condition).

Keywords: Fitness app · Fitness user · Sport habits · Virtual fitness training

1 Introduction

The use of the internet continues to grow worldwide, and to a greater extent in recent years. Of particular note is internet access via mobile phones, which reached 5.2 billion

B. Sañudo Corrales and J. García-Fernández (Eds.): Tapasconference 2020, LNBE, pp. 88–97, 2022.
https://doi.org/10.1007/978-3-030-92897-1_10

devices in 2020, representing a penetration of 67% worldwide [1]. This same platform states that there are a total of 4,140 million users of social networks, which corresponds to 53% of global penetration. In the world, 53.3% of all web traffic is carried out via mobile phones [2].

In recent years, sport has continued to evolve towards greater organization and professionalization in its strategies, resources, personnel, and business processes [3]. The sports industry, with an estimated global economic value of $805 billion worldwide, has a major economic and social impact [4]. The fitness sector provides its services with a commercial purpose [5], becoming a global phenomenon that has evolved and diversified a multitude of training techniques and business models aimed at improving people's health through the performance of qualified professionals [6]. The last decade has seen an exponential increase in the number of users in fitness centers from 119.53 million in 2009 to more than 183 million in 2018, distributed in more than 210,000 centers worldwide [7].

An important step to be developed by fitness centers and the sports sector in general, in order to continue growing, is the widespread application of digitalization. The changes produced by digitalization have also affected the administration and management of fitness sports centers, which trust their information systems to pre-configured standard packages such as Enterprise Resource Planning (ERP) [8]. These solutions are based on application software modules which help to manage the important parts of their organization, such as invoicing, financial management, sales, production, activities and reservations, human resources. Recently it can be seen how they include, within their standard, other more innovative ones such as gamification, internet, projects or document management, among others [9].

The sport and fitness fields are considered health-related industries, and the growth of this link in association with mobile applications (apps) is exponential [10]. According to Statista [11], 78% of fitness center users use fitness and sports training apps. A study developed by Knight et al. [12] determined the lack of evidence of useful apps for the improvement of health. This study suggests an opportunity to develop apps that have a clinical use for improving health. For this reason, an increase was observed in the number of health-related apps that appeared between 2012 and 2015 in the Spanish fitness sector [13].

Fitness centers are increasingly betting on implementing the use of mobile applications while increasing the offer of apps, with the aim of generating greater adherence to physical practice and loyalty to the center itself [14]. Because the use of apps and other digital platforms provide more segmented information on customers and services, it is possible to personalize the stakeholders and strategies, by generating new digital ecosystems [15]. In addition, it will also help to have a greater presence of the brand and generate greater awareness of it [16]. Joszt [17] reported an increase in adherence to physical activity practice through the use of health apps in the clinical context in patients receiving therapy.

Serdà-Ferrer et al. [18] found that the variables that most encouraged adherence to physical practice were motivation, satisfaction, experience and perception of change or improvement. The apps related to physical exercise allow the quantification of exercise and the control of the evolution of the practice of physical activity. These improvements or changes can thus lead to greater adherence to physical activity. In turn, a study noted that the incorporation in apps of techniques such as goal setting and intention formation could increase user participation and promote their continued use [19].

In a study on the efficacy of an app to promote pelvic floor exercise in women, Stephen [20] reported that the use of the app was shown to be insufficient to produce a change in individual behavior and health. Another study evaluating participants seeking to improve the adherence to a healthy lifestyle found that the social support generated by the app through teamwork increased adherence and commitment to healthy habits [21]. Finally, Wang et al. [22] conducted a systematic review of experimental studies that used apps or digital devices for weight loss in overweight or obese individuals. Authors suggested that, while considering the importance of technology, many other outcomes must be involved.

2 Objective

The aim of the current study was to analyze changes in the sports habits of participants, attributable to the use of the center's mobile application.

3 Method

3.1 Study Design

This study had a non-experimental, descriptive analytical cross-sectional design using retrospective evidence-based cohort sampling and sample accessibility.

3.2 Sample

The sample was composed of 1,285 subjects, belonging to 17 sports centers. The socio-demographic data (Table 1) consists of 55.0% female and 44.8% male, with an age of 36 to 50 years (43.1%), followed by those participants between 21 and 35 years (25.1%). In terms of educational level, the majority had university studies and vocational education with 39.4% and 32.1% respectively.

3.3 Instrument

The instrument used was composed of different sections. A first section included informative questions about the sports center they belong to and the use and promotion of apps, specifying whether they have actually downloaded the center's app, and how it is employed. The second section was related to technology and exercise and focused on the users' knowledge of the habit of using sports applications available in fitness sports centers [23]. The last section was formed by the socio-demographic variables (gender, age and level of studies) and the sports habits they had when they registered at the sports center, before and after the use of the app, and currently (weekly practice frequency, duration in the center each day, type of activity performed and level of fitness experience).

Table 1. Descriptive statistics of the sample.

Variable	Value	N	%
Gender	Male	576	44.8
	Female	707	55.0
	Other	2	0.2
Age (years)	< 20	35	2.7
	21 – 35	323	25.1
	36 – 50	556	43.1
	51 – 65	320	25.0
	> 65	51	4.1
Education level	No studies	8	0.6
	Elementary studies	93	7.2
	Vocational education	163	32.1
	University studies	506	39.4
	Postgraduate	214	20.7

3.4 Procedure

Four operators of an important franchise chain of fitness centers in Spain were contacted, explaining to them the objective of the study and their possible participation. The directors of the chains reviewed the survey, confirming their interest in participating. The data collection procedure was carried out by means of an on-line form developed with the Google "Forms" tool (Google LLC, Mountain View, California, USA), sent by the fitness centers themselves. They were responsible for the mass mailing to their databases, guaranteeing their users the confidentiality of the information, as well as voluntary participation. In the mail sent, the purpose and objective of the survey was explained to its customers, and they were asked for the utmost sincerity and honesty in the absence of correct or incorrect answers. The approximate time to complete the different items of the questionnaire was between 8–10 min. The period for completion of the questionnaire was between 26 April 2019 and 27 June 2019.

3.5 Data Analysis

Data analysis was performed using SPSS v.24.0 (IBM, Armonk, USA). Descriptive statistics were calculated for the qualitative variables, frequencies and percentages. On the other hand, contingency tables were made using the McNemar test and the Fisher Statistician, in order to analyse the possible differences in the change in the sports habits of the subjects upon registration and currently, and between those who downloaded the application and those who did not. The effect size was calculated using Cramer's V [24]. The significance level was a value of $p \leq 0.05$.

4 Results

Comparing the sports experience of the participants of the sample at the time of registration to the center and at present (Table 2), the results indicated that the regularity the users went to the center was between 3 and 4 days per week, being higher at the time of registration than at the time of the study. In terms of time spent at the center, it was observed that most users spent between 60 and 90 min, with the time spent at the center increasing slightly up to the time of study.

Table 2. Values on the sports experience of the exhibition when it was registered at the center and currently.

Variable	Registration date	Currently	X^2	*p-value*	V Cramer
	N(%)	N(%)			
Times per week she/he went/is going to the center					
One day or none	41(3.2)	133(10.4)	67.6(6)	0.001*	0.372
1 - 2 days	297(23.1)	247(19.2)			
3 - 4 days	745(58.0)	674(52.5)			
5 days or more	202(15.7)	231(18.0)			
Time of permanence each day that he/she went/is going to the center					
Less than 30 min	16(1.2)	57(4.4)	52.5(6)	0.001*	0.478
30 – 60 min	477(37.1)	418(32.5)			
60 – 90 min	670(52.1)	650(50.6)			
More than 90 min	122(9.5)	160(12.5)			
Fitness experience level					
Low level	645(50.2)	266(20.7)	382.2(3)	0.001*	0.421
Medium level	549(42.7)	835(65.0)			
High level	91(7.1)	184(14.3)			
Type of activities					
Group training	587(45.7)	615(47.9)	376.1(1)	0.001*	0.544
Fitness room	517(40.2)	597(46.5)	317.9(1)	0.001*	0.500
Fitness Cardio	436(33.9)	564(43.9)	268.1(1)	0.001*	0.459
Swimming	279(21.7)	286(22.3)	458.3(1)	0.001*	0.600
Aqua fitness	44(3.4)	54(4.2)	234.9(1)	0.001*	0.430
Other	36(2.8)	39(3.0)	114.3(1)	0.001*	0.300

Note. P value; * $p \leq 0.001$

Related to the level of fitness experience, there is an increase in the level of fitness experience as the user uses and stays in the center, highlighting the increase of 22.3% with a medium level and 7.2% with a high level. Finally, in terms of the type of activities, the most practiced were group training activities, followed by fitness room and fitness cardio, both when registering and at present. An increase in practice was observed in all the activities at the time of study with respect to the initial moment.

The results showed statistically significant differences in all the variables, improving the sport habits and their practice within the sport centers in all the users that have app ($p \leq 0.05$). The effect of the variables was moderate in all of them except in the swimming activity, which was high.

Table 3. Values on the sports experience of the current sample based on the app download.

Variable	Registration date	Currently	X^2	*p-value*	V Cramer
	N(%)	N(%)			
Times per week she/he went/is going to the center					
One day or none	45(6.8)	88(14.0)	25.8(3)	0.001*	0.142
1 - 2 days	114(17.4)	133(21.2)			
3 - 4 days	380(57.8)	294(46.8)			
5 days or more	118(18.0)	113(18.0)			
Time of permanence each day that he/she went/is going to the center					
Less than 30 min	19(2.9)	38(6.1)	19.6(3)	0.001*	0.124
30 – 60 min	191(29.1)	227(36.1)			
60 – 90 min	367(55.9)	283(45.1)			
More than 90 min	80(12.2)	80(12.7)			
Fitness experience level					
Low level	102(15.5)	164(26.1)	21.9(2)	0.001*	0.131
Medium level	455(69.3)	380(60.5)			
High level	100(15.2)	84(13.4)			
Type of activities you used to practice					
Group training	383(58.6)	232(37.1)	58.1(1)	0.001*	0.215
Fitness room	321(49.1)	276(44.2)	3.1(1)	0.082	0.049
Fitness Cardio	300(45.9)	264(42.2)	1.7(1)	0.196	0.037
Swimming	104(15.9)	182(29.1)	32.2(1)	0.001*	0.159
Aquafitness	35(5.4)	19(3.0)	4.2(1)	0.051	0.057
Other	17(2.6)	22(3.5)	0.9(1)	0.416	0.027

Note. P value; * $p \leq 0.001$

Analyzing the results of the sports habits based on the users' app downloads (Table 3) showed that the users who downloaded the app had a higher frequency of attendance and stayed longer in the sports center than those users who did not download the app. The user experience level was also higher in those who downloaded the app. The results of the type of activities practiced indicated that, in general, the users who downloaded the app had a greater participation in all the different activities, except swimming.

The results showed statistically significant differences in frequency, duration in the center, level of fitness experience and in the directed activities and swimming ($p \leq 0.05$). The effect of the variables was small in those that showed statistically significant differences, while the variables that did not obtain differences in effects were irrelevant according to both user groups.

5 Discussion and Conclusions

The results between the sports experience of the participants when registering and now, as well as before the use of the application and after, showed statistically significant differences in the frequency of attendance per week and the time spent in the fitness center. These factors were determinant for the adherence to the fitness center, coinciding with a study on overweight adults, where statistically significant differences were observed between before using the application and after [25]. However, Direito et al. [26] found favorable results in young people to increase physical activity, although the results did not show statistically significant differences.

Regarding the number of times users came to the center, a statistically significant difference was found between users who downloaded the application versus those who did not ($p \leq 0.001$); the latter reducing the attendance by one day or none, and between 1 and 2 days, increasing the attendance by 3 to 4 days and 5 or more days. In comparison to the time spent per training session, we also found statistically significant differences in the time spent in the fitness center among the users who downloaded the app ($p \leq 0.001$). There were decreases in the time of permanence of less than 30 min, and from 30 to 60 min, increasing the permanence between 60 and 90 min (10,8%) and being very similar the one of more than 90 min. These results coincide with studies by Allen [27], Du et al. [21], Jacobs et al. [28], Silveira et al. [29] or Voth et al. [30], which conclude that the use of technology favors adherence to physical practice.

As for the level of fitness of the study participants, we observed statistically significant changes in all values, which indicated that the use of the application improved knowledge about exercise performance and one's own experience ($p \leq 0.001$). After using the application, the low level of fitness decreased by 10.6%, the medium level increased by 8.8% and the high level increased by 1.8%, compared to those who did not download the application. These results have already been shown in previous studies such as that of Ritterband et al. [31] who stated that the use of the website or mobile applications could lead to a change in behavior due to different mechanisms such as improved knowledge of both the results and the technical level, increased experience and motivation to achieve.

In relation to the type of activity carried out, we also found differences between those who downloaded the application and those who did not. There was an increase in those who downloaded the application compared to those who did not, of 21.5%, in the carrying out of group training activities, 4.9% in the training in the fitness room, 3.7% in the fitness cardio area, and 2.4% in aqua fitness, reducing on the other hand the practice of swimming and other activities. These data are similar to the study carried out by Estrada-Marcén et al. [23] where users of fitness technologies had a higher percentage in fitness room and group training activities and less in aquatic, high intensity or wellness activities.

5.1 Conclusions

The main findings of the study show a general improvement in the sports habits of users from the moment of registration, especially in the time spent training and the level of fitness experience. In fact, the effect on changes in the various habits analyzed at the time of registration with respect to the current time is moderate.

Comparing the differences between users who downloaded and used the mobile application and those who did not, the former presented higher percentages in sports habits, increasing their frequency of weekly attendance, spending more time in training and having a higher level of fitness experience than those who did not download the application. However, the effect produced in these changes when downloading the app was small.

Finally, analyzing the type of activities carried out, it was observed that there was a general increase in the participation of the services offered from the moment of registration by the users. On the part of those users who downloaded and used the center's mobile application, differences were also observed between those who did not, as they carried out more group training activities and less swimming, also observing a trend towards significant use of the fitness room.

5.2 Limitations and Future Research

The main limitation can be found in the selection of the participating sports centers, which was made for the convenience of accessibility to the sample. Due to this aspect, we understand that the results obtained could not be generalized. On the other hand, the variables of the study do not contemplate the analysis of personal data and its possible influence on the use of technology according to age, gender or socio-demographic profile, although it is true that this was not the main objective of this study.

Once this study has been carried out, various future research lines are considered which will provide continuity and depth to the study. It would be considered interesting to be able to study the influence of technology and the use of mobile apps according to the business model of the fitness center. On the other hand, it would also be very interesting to be able to analyse the implementation method of the mobile app of each center, evaluating its impact on users and its differences according to the aforementioned business model.

Another line of research that could improve this study would be the analysis of the influence and use of technology according to the customer profile, taking into account age, gender or socio-demographic condition. In this way, we would know its influence according to user segmentation and how it affects the different variables analysed, considering the different generations and how they are linked to the use and acceptance of technology.

5.3 Practical Implications

This study would be of great interest to the fitness sector and the community To be able to develop longitudinal studies, which show results on different time scales and can be compared with each other, and in this way obtain a broader view of the influence of the use of technology on centers and users.

As recommendations that may be of interest for sports management and the use of technology in fitness centers, it is worth commenting that the digital revolution is bringing about a change of era in our social, economic, industrial, educational environment, etc. and the sports and fitness sector is also affected by technological development.

The digital transformation of sports facilities does not only affect a first layer of process management, protocols or tools which facilitate their development, but it also affects a second layer of consumption, practice, relationship and user experience. We must be able to provide the best experience at all times and in all aspects of the provision of our service and in all channels, also having, if possible, a mobile application because of its positive effect on users' sporting habits.

References

1. https://datareportal.com/reports/digital-2020-october-global-statshot
2. https://ditrendia.es/informe-mobile-2020/
3. Xiao, X., et al.: Sports digitalization: an overview and a research agenda. In: Proceedings of the 38th International Conference on Information Systems 2018, pp. 4377–4397. AIC (2018)
4. https://www.pwc.com/CISwearables
5. Riseth, L., Nøst, T.H., Nilsen, T.I., Steinsbekk, A.: Long-term members' use of fitness centers: a qualitative study. BMC Sports Sci. Med. Rehabil. **11**(2), 1–9 (2019). https://doi.org/10.1186/s13102-019-0114-z
6. Andreasson, J., Johansson, T.: Glocalised fitness: the franchising of a physical movement, fitness professionalism and gender. Leisure/Loisir **42**(3), 301–321 (2018). https://doi.org/10.1080/14927713.2018.1535910
7. https://www.ihrsa.org/publications/the-2019-ihrsa-global-report/
8. Valcarce, M.: Tecnología y gestión de entidades deportivas. Universidad Internacional de Valencia (2019)
9. Haddara, M.: ERP selection: the SMART way. Procedia Technol. **16**, 394–403 (2014). https://doi.org/10.1016/j.protcy.2014.10.105
10. Sañudo, B.: El caso de las wearables technologies en la actividad física y el deporte. In: García-Fernández, J. (Ed.) Emprendiendo en el sector deportivo. De la teoría a la práctica, pp. 73–84. Aranzadi Thomson Reuters (2016)
11. https://es.statista.com/estadisticas/942090/uso-de-aplicaciones-en-espana/
12. Knight, E., Stuckey, M.I., Prapavessis, H., Petrella, R.J.: Public health guidelines for physical activity: Is there an app for that? A review of android and apple app stores. JMIR Mhealth Uhealth **3**, e43 (2015). https://doi.org/10.2196/mhealth.4003

13. Valcarce, M., Miñambres, T., Hernández, C.C.: Impacto económico en las empresas que ofrecen aplicaciones móviles en el sector del fitness en España. In: Proceedings of the IV Congreso Iberoamericano de Economía del Deporte 2016, pp. 175–180. SEED (2016)
14. Valcarce, M., Díez, C.: Influencia de una app en la adherencia a la práctica deportiva: protocolo de estudio. Revista de Educación Motricidad e Investigación. **11**, 16–34 (2018)
15. Gruettner, A.: What we know and what we do not know about digital technologies in the sports industry. In: Proceedings of the 25th Americas Conference on Information Systems 2019, pp. 1–10. AMCIS (2019)
16. Selman, H.: Marketing digital. Ibukku (2017)
17. Joszt, L.: Things About Wearable Technology in Healthcare. The American Journal of Managed Care (2018)
18. Serdà-Ferrer, B.C., Valle, A.D., Marcos-Gragera, R.: The adherence to physical exercise in a group of prostate cancer: an integrated model to improve the quality of life. Psychosoc. Interv. **21**(1), 29–40 (2012). https://doi.org/10.5093/in2012v21n1a4
19. Helander, E., Kaipainen, K., Korhonen, I., Wansink, B.: Factors related to sustained use of a free mobile app for dietary self-monitoring with photography and peer feedback: retrospective cohort study. J. Med. Internet Res. **16**(4), e109 (2014). https://doi.org/10.2196/jmir.3084
20. Stephen, C.: Adherence to pelvic floor muscle exercises and the role of Smart phone apps, University of Aberdeen (2015)
21. Du, H., Venkatakrishnan, A., Youngblood, G.M., Ram, A., Pirolli, P.: A group-based mobile application to increase adherence in exercise and nutrition programs: a factorial design feasibility study. JMIR. mHealth. uHealth. **4**(1), e4 (2016). https://doi.org/10.2196/mhealth.4900
22. Wang, E., Abrahamson, K., Liu, P.J., Ahmed, A.: Can mobile technology improve weight loss in overweight adults? A systematic review. West. J. Nurs. Res. **42**(9), 747–759 (2020). https://doi.org/10.1177/0193945919888224
23. Estrada-Marcén, N., Simon-Grima, J., Sanchez-Bermudez, J., Casterad-Sera, J.: Use of fitness gadgets by gym users. Retos **38**, 26–32 (2020). https://doi.org/10.47197/retos.v38i38.73108
24. Dominguez-Lara, S.: Effect size, a quick guide. Educación Médica **19**, 251–254 (2018). https://doi.org/10.1016/j.edumed.2017.07.002
25. Watson, A., Bickmore, T., Cange, A., Kulshreshtha, A., Kvedar, J.: An internet-based virtual coach to promote physical activity adherence in overweight adults: randomized controlled trial. J. Med. Internet Res. **14**(1), e1 (2012). https://doi.org/10.2196/jmir.1629
26. Direito, A., Jiang, Y., Whittaker, R., Maddison, R.: Apps for IMproving FITness and increasing physical activity among young people: the AIMFIT pragmatic randomized controlled trial. J. Med. Internet Res. **17**(8), e210 (2015). https://doi.org/10.2196/jmir.4568
27. Allen, V.T.: Automated Technology Based Behavioural Intervention to Promote Exercise Adherence: A Pilot Study to Ascertain Efficacy. University of Waikato (2015)
28. Jacobs, S., Radnitz, C., Hildebrandt, T.: Adherence as a predictor of weight loss in a commonly used smartphone application. Obes. Res. Clin. Pract. **11**(2), 206–214 (2017). https://doi.org/10.1016/j.orcp.2016.05.001
29. Silveira, P., Van het Reve, E., Daniel, F., Casati, F., De Bruin, E.D.: Motivating and assisting physical exercise in independently living older adults: a pilot study. Int. J. Med. Inform. **82**(5), 325–334 (2013). https://doi.org/10.1016/j.ijmedinf.2012.11.015
30. Voth, E.C., Oelke, N.D., Jung, M.E.: A theory-based exercise app to enhance exercise adherence: a pilot study. JMIR Mhealth Uhealth **4**(2), e62 (2016). https://doi.org/10.2196/mhealth.4997
31. Ritterband, L.M., Thorndike, F.P., Cox, D.J., Kovatchev, B.P., Gonder-Frederick, L.A.: A behavior change model for internet interventions. Ann. Behav. Med. **38**(1), 18–27 (2009). https://doi.org/10.1007/s12160-009-9133-4

Applying Machine Learning to Estimate Osteoporosis Risk Based on Compliance with WHO Guidelines for Physical Activity in Postmenopausal Women

Data from the 2017–18 National Health and Nutritional Examination Surveys (NHANES)

Horacio Sanchez-Trigo[1](✉), Emilio Molina[1], Sergio Tejero[2](✉), and Borja Sañudo[1]

[1] Department of Physical Education and Sports, Faculty of Educational Sciences, Universidad de Sevilla, 41013 Seville, Spain
fstrigo@us.es
[2] Hospital Universitario Virgen del Rocío, 41013 Sevilla, Spain

Abstract. Introduction: Contemporary statistical approaches coming from Data Mining and Machine Learning can help us identify associations between data and build predictions models. Objective: To predict osteoporosis risk based on compliance to WHO guidelines for physical activity in postmenopausal women. Method: Data from 917 postmenopausal women from the 2017–18 National Health and Nutritional Examination Surveys (NHANES) was analyzed using machine learning algorithms with the aim of estimating osteoporosis risk based on a supervised learning perspective. Input data for the prediction algorithm included physical activity data collected via questionnaire. Results: From the information obtained by applying a decision tree algorithm, osteoporosis risk was estimated at 27,8% for postmenopausal women complying to WHO guidelines for physical activity, versus a 31,5% risk for non-compliers. Therefore, adherence to WHO recommendations could be associated to an osteoporosis risk reduction of 3,7%. Conclusion: Machine Learning and Data Mining algorithms can help us identify key factors in preventing osteoporosis. Our analysis shows that complying to WHO guidelines for physical activity could have a significant impact on bone health.

Keywords: Osteoporosis · Physical activity · WHO · Machine learning · Data mining

1 Introduction

Osteoporosis is defined as a skeletal disease characterized by reduced bone mass and disruption of bone architecture, leading to increased bone fragility and fracture risk [1]. Osteoporosis is highly prevalent worldwide, affecting approximately 27.5 million adults in the European Union [2] and 10.2 million people in the United States [3]. Furthermore, in average one in three women and one in five men over the age of 50

B. Sañudo Corrales and J. García-Fernández (Eds.): Tapasconference 2020, LNBE, pp. 98–106, 2022.
https://doi.org/10.1007/978-3-030-92897-1_11

years suffers an osteoporotic fracture [4]. As a consequence, the economic burden of osteoporosis is very high. Approximately €37 billion are spent in Europe annually to cover the costs associated with osteoporotic fractures hospitalization and aftercare [2]. In Spain approximately 2.5 million people suffer from osteoporosis, generating health care costs of €3.5 billion per year [5]. Moreover, due to the population ageing these costs are expected to double by 2050 [6]. Osteoporosis is therefore a major public health concern worldwide, so it is utter importance that effective osteoporosis prevention programs are developed.

Regular exercise is one of the strategies that may effectively contribute in preventing osteoporosis. Research shows that different exercise modalities (e.g., aerobic exercise and strength training) have a positive impact on the bone health of both pre- and post-menopausal women [7]. As living tissues, bones have the capacity to adapt to the stimuli they are subjected to. Thus, bone adapt to physical activity by increasing their density and improving their microarchitecture with the aim of resisting biomechanical demands more efficiently [8]. In order to increase bone mineral density (BMD), bone mass and bone strength, physical activity stimulates osteogenic pathways and the activities of osteoblasts and osteocytes. Besides osteo-clastogenesis and bone resorption are downregulated [9]. Bone mass growth requires that bone tissue is stimulated with loads exceeding a certain threshold, as the osteogenic response is function of the amount of mechanical force applied on the skeletal system [10]. Evidence shows that the regular practice of high-impact and weight-bearing exercise could exceed such threshold and contribute positively to bone health, nevertheless, knowledge of the role that physical activity plays in preventing osteoporosis is still incomplete [11]. It is therefore of interest to further research the relationships between physical activity and osteoporosis, in order to develop effective and safe protocols.

Data mining techniques, also known as 'knowledge discovery in databases', allow discovering novel and potentially useful information from large amounts of data and hidden relationships among these data [12]. These patterns and trends in databases uncover new paths for further analyses and permit building predictive models. Machine learning algorithms form the core of data mining [13]. Machine learning can be defined as computational methods that aim to make accurate predictions based on past information [14]. In other words, these contemporary statistical approaches can help us identify associations between data and build predictions models. Machine learning has been intensively used in many fields such as marketing, banking, quality control, etc. It is also becoming increasingly popular in healthcare, as it holds great potential to improve health systems by identifying best practices that improve care and reduce costs [15].

In this study machine learning is employed to analyze the mediating role of physical activity in osteoporosis. Specifically, machine learning algorithms are used to assess whether compliance with certain activity levels could be associated to a lower osteoporosis incidence, and therefore identify which practices might be more effective in preventing this highly prevalent disease. To our knowledge no previous studies have employed machine learning algorithms to estimate osteoporosis risk based on physical activity data.

2 Objective

The aim of this study was to identify which of the physical activity variables included in the National Health and Nutrition Examination Survey (NHANES) 2017–18 has the greater association with osteoporosis diagnosis, in order to identify best practices and potential lifestyle modifications to reduce osteoporosis risk among postmenopausal women.

3 Method

3.1 Data collection

This is a cross-sectional study using data from the National Health and Nutrition Examination Survey (NHANES) 2017–18. A comprehensive descriptions of the NHANES 2017–18 study design and procedures are available elsewhere [16]. To conduct this study, femoral neck BMD and physical activity data were extracted from the NHANES 2017–18 database and processed as follows:

- Femoral neck BMD values, which were measured by dual-energy X-ray absorptiometry [17], were converted in this study to T-Score using updated reference data on femur bone mineral levels of noninstitutionalized US adults differentiating by sex and ethnicity [18]. According to the criteria defined by the World Health Organization (WHO), osteoporosis is diagnosed in postmenopausal women aged 50 years or more when the T-score lies below –2.5 [19]. Thus, subjects who met these criteria were identified.
- Physical activity data, which were collected via the Global Physical Activity Questionnaire (GPAQ that include questions concerning daily activities, leisure time activities, and sedentary activities [20–22]) were used in this study. Data on weekly moderate-intensity physical activity and vigorous physical activity, as well as daily sedentary time were extracted from the questionnaires.

Using these data, compliance with WHO recommendations on physical activity was verified. The WHO recommends adults to "*do at least 150 min of moderate-intensity aerobic physical activity throughout the week or do at least 75 min of vigorous-intensity aerobic physical activity throughout the week or an equivalent combination of moderate- and vigorous-intensity activity*" [23]. Therefore, to test if the subjects complied to these recommendations, the sum of the weekly moderate physical activity plus twice the weekly vigorous physical activity was calculated. If this sum lies above 150 min, then the WHO recommendations are met.

IF(weekly moderate PA + 2 * weekly vigorous PA)>150 min
THEN recommendations met
ELSE recommendations not met

This test contemplates all possibilities to meet the WHO recommendations, that is, at least 150 min of moderate-intensity aerobic physical activity throughout the week, or at least 75 min of vigorous-intensity aerobic physical activity throughout the week, or an equivalent combination of moderate- and vigorous-intensity activity.

3.2 Statistical Analysis

The statistical analysis framework carried out in this study is based on a decision tree algorithm. A decision tree is a data mining technique that has the advantage of ease of interpretation and visualization [15]. The goal of a decision tree is creating a model that predicts the value of a target variable based on several input variables. A decision tree model is composed of nodes (points of decision making), branches (representing decision possibilities based on the values of the input variables) and leaves (representing a classification outcome) as shown in Fig. 1.

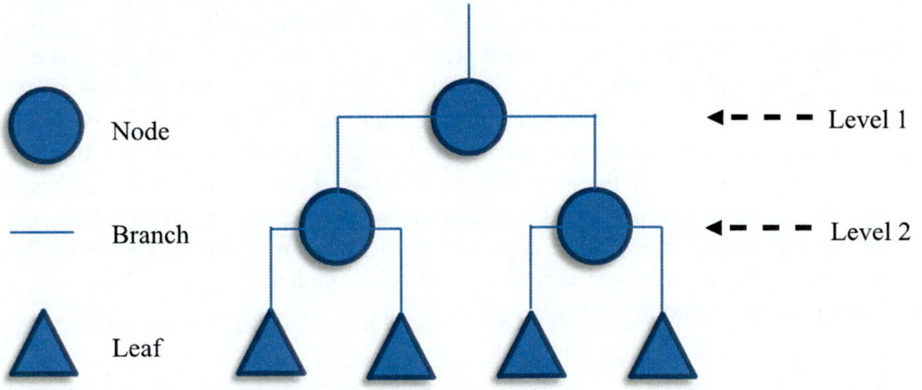

Fig. 1. Two levels decision tree model.

Therefore, a decision tree model is structured as a hierarchical set of conditions. An instance (subject) is classified by following the path of verified conditions node by node from the root of the tree until a leaf is reached. The depth of a tree is defined by the number of levels in this decision-making process. At every level, the decision tree considers all input variables and chooses the split which results in most homogeneous sub-nodes, thus, the variable situated at level 1 is the most decisive input variable in predicting the target variable [24]. The lower a variable appears in the tree, the lower its power to classify instances, therefore, a decision tree is one of the quickest algorithms to identify the most significant relationships between two or more variables, making it a powerful tool when trying to quickly identify which features have a higher impact on a certain outcome [25]. In our case, a decision tree was used to identify which of the variables extracted from the physical activity questionnaire have a higher relationship with the diagnosis of osteoporosis (T-score below –2.5).

- Input/Predictive variables: weekly vigorous physical activity [minutes], weekly moderate physical activity [minutes], compliance with WHO physical activity recommendations [yes/no], daily sedentary time [minutes].

- Output/Predicted variable: osteoporosis diagnosis [yes/no].

To perform this analysis the well-known WEKA (Waikato Environment for Knowledge Analysis) Data Mining software was employed [26].

The decision tree implemented in WEKA is a classifier called J48. The J48 classifier is based on the C4.5 algorithm, which generates pruned or unpruned decision trees [27].

4 Results

The NHANES 2017–2018 database contains a total of 9254 entries, of which 2286 include femoral neck BMD measurements. Out of these subjects, 33 were excluded due to lacking data in the physical activity questionnaire. Only female participants were selected, thus excluding 1182 male participants. 116 subjects from 80 years old and above were also excluded, as individuals over 80 are top coded at 80 years of age in the NHANES 2017–2018 database, so their precise age is missing [28]. Next, 38 outliers were identified and eliminated, so finally 917 subjects were included in the analysis (Fig. 2). Data from these 917 subjects was fed into the J48 algorithm in Weka. The resulting decision tree had the variable "compliance with WHO physical activity recommendations [yes/no]" in the first node. As mentioned above, the variable associated to the first decision node is the most significant one to predict the analyzed outcome. To further quantify the importance of observing the WHO physical activity recommendations, a comparison between compliers and non-compliers was carried out with the following results (Table 1):

Table 1. Comparison between compliers and non-compliers with the WHO physical activity recommendations

	Compliers	Non-compliers
N	396	521
Age (mean ± SD)	62 ± 7	63 ± 8
T-score (mean ± SD)	– 1.90 ± 0’99	– 1.95 ± 1.05
Osteoporosis (n)	110	164
Osteoporosis prevalence	27.8%	31.5%

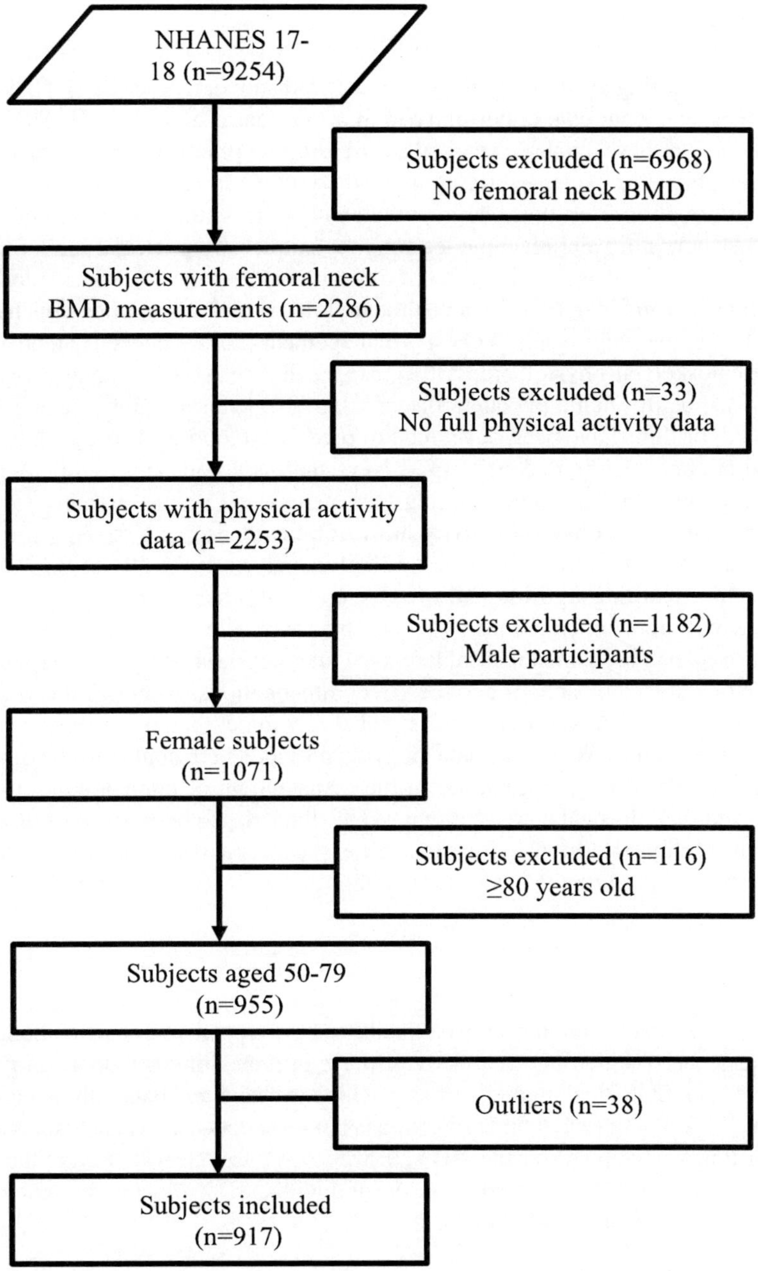

Fig. 2. Flow chart of inclusion and exclusion criteria

5 Discussion

There is an increasing availability of electronic healthcare databases like NHANES, and this provides an exceptional opportunity to improve health services [29]. The information contained in these databases may allow developing practical applications for early disease detection and patient care, among others uses. In order to efficiently extract valuable information from these large complex datasets, computational techniques are essential. Data mining and machine learning techniques are powerful tools for knowledge discovery in databases and for identifying patterns in data [12]. Some authors claim that machine learning can transform healthcare [29]. Promising applications have been developed in varied fields such as obesity management [30], epilepsy [31], heart failure [32], musculoskeletal physiotherapy [33], cancer diagnosis [34] among many others. Osteoporosis, being such a prevalent disease, has also been researched using machine learning. Algorithms have been developed to predict osteoporosis risk [35], bone fractures [36] or abnormal bone density [37]. Nevertheless, to our knowledge no previous studies have employed machine learning for osteoporosis risk assessment focused on physical activity data. Physical activity has a solid evidence as a strategy for the prevention and treatment of low bone mass [37]. Our analysis of the NHANES 2017–2018 with a machine learning algorithm allowed us to identify that complying with the WHO physical activity recommendations could be associated with a 3.7% reduction in osteoporosis disease risk in premenopausal women. Our analysis shows that osteoporosis risk could also be reduced by adhering to these recommendations. Although it is beyond the scope of the current study, observing the global recommendations on physical activity for health published by WHO, the data analyses provided here could also be of any help assessing the role of physical activity in the management of clinical conditions, as it was observed that physical activity might reduce the risk of coronary heart disease and stroke, diabetes, hypertension, colon cancer, breast cancer and depression [23]; however, some questions regarding the outcomes mediating these associations remain.

6 Conclusion

Evidence shows that different exercise modalities have a positive impact on bone health, nevertheless, the role that physical activity plays in preventing osteoporosis is not yet conclusive [11]. Statistical methods from machine learning and data mining can help in identifying best practices and find associations between data. In this analysis, a decision tree algorithm was applied on the NHANES 2017–18 to detect that compliance with the WHO recommendations could be associated with a 3.7% reduction in osteoporosis disease risk in premenopausal women.

References

1. NIH Consensus Development Panel on Osteoporosis Prevention, D., Therapy: Osteoporosis Prevention, Diagnosis, and Therapy. JAMA **285**(6), 785–795 (2001). doi:https://doi.org/10.1001/jama.285.6.785

2. Hernlund, E., et al.: Osteoporosis in the European union: medical management, epidemiology and economic burden. Arch. Osteop. **8**(1–2), 1–115 (2013). https://doi.org/10.1007/s11657-013-0136-1
3. Wright, N.C., et al.: The recent prevalence of osteoporosis and low bone mass in the United States based on bone mineral density at the femoral neck or lumbar spine. J. Bone Mineral Res. Official J. Am. Soc. Bone Mineral Res. **29**(11), 2520–2526 (2014). https://doi.org/10.1002/jbmr.2269
4. Sozen, T., Ozisik, L., Basaran, N.C.: An overview and management of osteoporosis. Eur. J. Rheumatol. **4**(1), 46–56 (2017). https://doi.org/10.5152/eurjrheum.2016.048
5. Svedbom, A., et al.: Osteoporosis in the European union: a compendium of country-specific reports. Arch. Osteop. **8**(1–2), 1–218 (2013). https://doi.org/10.1007/s11657-013-0137-0
6. Compston, J.: Osteoporosis: social and economic impact. Radiol. Clin. North Am. **48**(3), 477–482 (2010). https://doi.org/10.1016/j.rcl.2010.02.010
7. Pedersen, B.K., Saltin, B.: Exercise as medicine - evidence for prescribing exercise as therapy in 26 different chronic diseases. Scand. J. Med. Sci. Sports **25**(Suppl 3), 1–72 (2015). https://doi.org/10.1111/sms.12581
8. Borer, K.T.: Physical activity in the prevention and amelioration of osteoporosis in Women : interaction of mechanical, hormonal and dietary factors. Sports Med. **35**(9), 779–830 (2005). https://doi.org/10.2165/00007256-200535090-00004
9. Yuan, Y., et al.: The roles of exercise in bone remodeling and in prevention and treatment of osteoporosis. Prog. Biophys. Mol. Biol. **122**(2), 122–130 (2016). https://doi.org/10.1016/j.pbiomolbio.2015.11.005
10. Sanudo, B., et al.: A systematic review of the exercise effect on bone health: the importance of assessing mechanical loading in perimenopausal and postmenopausal women. Menopause **24**(10), 1208–1216 (2017). https://doi.org/10.1097/GME.0000000000000872
11. Kai, M.C., Anderson, M., Lau, E.: Exercise interventions: defusing the world's osteoporosis time bomb. Bull. World Health Organ. **81**, 827–830 (2003)
12. Baker, R.: Data mining for education. Int. Encycl. Educ. **7**(3), 112–118 (2010)
13. Mannila, H.: Data mining: machine learning, statistics, and databases. In: Proceedings of 8th International Conference on Scientific and Statistical Data Base Management, pp. 2–9. IEEE (1996)
14. Mohri, M., Rostamizadeh, A., Talwalkar, A.: Foundations of Machine Learning. MIT Press, Cambridge (2018)
15. Koh, H.C., Tan, G.: Data mining applications in healthcare. J. Healthcare Inf. Manage. **19**(2), 65 (2011)
16. https://wwwn.cdc.gov/nchs/nhanes/continuousnhanes/overview.aspx?BeginYear=2017
17. https://wwwn.cdc.gov/Nchs/Nhanes/2017-2018/DXXFEM_J.htm
18. Looker, A.C., Borrud, L.G., Hughes, J.P., Fan, B., Shepherd, J.A.: Lumbar spine and proximal femur bone mineral density, bone mineral content, and bone area: United States, 2005–2008. Vita. Health Statist. Ser.11, Data Natl. Health Surv. **251**, 1–132 (2012)
19. Kanis, J.A., Melton, L.J., III., Christiansen, C., Johnston, C.C., Khaltaev, N.: The diagnosis of osteoporosis. J. Bone Mineral Res. **9**(8), 1137–1141 (1994)
20. https://wwwn.cdc.gov/Nchs/Nhanes/2017-2018/PAQ_J.htm
21. Armstrong, T., Bull, F.: Development of the world health organization global physical activity questionnaire (GPAQ). J. Public Health **14**(2), 66–70 (2006). https://doi.org/10.1007/s10389-006-0024-x
22. Cleland, C.L., Hunter, R.F., Kee, F., Cupples, M.E., Sallis, J.F., Tully, M.A.: Validity of the global physical activity questionnaire (GPAQ) in assessing levels and change in moderate-vigorous physical activity and sedentary behaviour. BMC Public Health **14**(1), 1255 (2014). https://doi.org/10.1186/1471-2458-14-1255

23. WHO: Global recommendations on physical activity for health. World Health Organization (2010)
24. Nilculaescu, O.: Classifying data with decision trees. XRDS: Crossroads ACM Mag. Stud. **24**(4), 55–57 (2018). https://doi.org/10.1145/3220677
25. Dileep Kumar, G.: Tree-Based Modeling Techniques. In: Dileep Kumar, G. (ed.) Machine Learning Techniques for Improved Business Analytics, pp. 1–18. IGI Global (2019). https://doi.org/10.4018/978-1-5225-3534-8.ch001
26. Hall, M., Frank, E., Holmes, G., Pfahringer, B., Reutemann, P., Witten, I.H.: The WEKA data mining software: an update. ACM SIGKDD Explor. Newslett. **11**(1), 10–18 (2009)
27. Quinlan, J.R.: Learning decision tree classifiers. ACM Comput. Surv. (CSUR) **28**(1), 71–72 (1996)
28. Wiens, J., Shenoy, E.S.: Machine learning for healthcare: on the verge of a major shift in healthcare epidemiology. Clin. Infect. Diseas. **66**(1), 149–153 (2017). https://doi.org/10.1093/cid/cix731
29. DeGregory, K.W., et al.: A review of machine learning in obesity. Obesity Rev. **19**(5), 668–685 (2018). https://doi.org/10.1111/obr.12667
30. Abbasi, B., Goldenholz, D.M.: Machine learning applications in epilepsy. Epilepsia **60**(10), 2037–2047 (2019). https://doi.org/10.1111/epi.16333
31. Awan, S.E., Sohel, F., Sanfilippo, F.M., Bennamoun, M., Dwivedi, G.: Machine learning in heart failure: ready for prime time. Current Opin. Cardiol. **33**(2), 190–195 (2018). https://doi.org/10.1097/hco.0000000000000491
32. Tack, C.: Artificial intelligence and machine learning|applications in musculoskeletal physiotherapy. Musculoskelet. Sci. Pract. **39**, 164–169 (2019). https://doi.org/10.1016/j.msksp.2018.11.012
33. Wong, D., Yip, S.: Machine learning classifies cancer. Nature **555**(7697), 446–447 (2018). https://doi.org/10.1038/d41586-018-02881-7
34. Yoo, T.K., et al.: Osteoporosis risk prediction for bone mineral density assessment of postmenopausal women using machine learning. Yonsei. Med. J. **54**(6), 1321–1330 (2013)
35. Ferizi, U., et al.: Artificial intelligence applied to osteoporosis: a performance comparison of machine learning algorithms in predicting fragility fractures from MRI Data. J. Magn. Reson. Imaging **49**(4), 1029–1038 (2019). https://doi.org/10.1002/jmri.26280
36. Lee, S., Choe, E., Kang, H., Yoon, J., Kim, H.: The exploration of feature extraction and machine learning for predicting bone density from simple spine X-ray images in a Korean population. Skelet. Radiol. **49**(4), 613–618 (2019). https://doi.org/10.1007/s00256-019-03342-6
37. Weaver, C.M., et al.: The national osteoporosis foundation's position statement on peak bone mass development and lifestyle factors: a systematic review and implementation recommendations. Osteop. Int. **27**(4), 1281–1386 (2016). https://doi.org/10.1007/s00198-015-3440-3

Author Index

B. Sañudo Corrales and J. García-Fernández (Eds.): Tapasconference 2020, LNBE , pp. 107–108, 2022.
https://doi.org/10.1007/978-3-030-92897-1